MD Banian

Managing Design

MD Baniamin Sarder

Managing Design Knowledge

Ontology Modeling in Product & Process Design

VDM Verlag Dr. Müller

Copyright © 2007 VDM Verlag Dr. Müller e. K. and licensors
All rights reserved. Saarbrücken 2007
Contact: info@vdm-verlag.de
Cover image: www.purestockx.com
Publisher: VDM Verlag Dr. Müller e. K., Dudweiler Landstr. 125 a, 66123 Saarbrücken, Germany
Produced by: Lightning Source Inc., La Vergne, Tennessee/USA
          Lightning Source UK Ltd., Milton Keynes, UK

Copyright © 2007 VDM Verlag Dr. Müller e. K. und Lizenzgeber
Alle Rechte vorbehalten. Saarbrücken 2007
Kontakt: info@vdm-verlag.de
Coverbild: www.purestockx.com
Verlag: VDM Verlag Dr. Müller e. K., Dudweiler Landstr. 125 a, 66123 Saarbrücken, Deutschland
Herstellung: Lightning Source Inc., La Vergne, Tennessee/USA
          Lightning Source UK Ltd., Milton Keynes, UK

ISBN: 978-3-8364-2837-8

DEDICATION

I would like to dedicate this book to my family members that include my wife Shimu,

two daughters Mariah and Sarah and also to my late parents.

ACKNOWLEDGEMENTS

I would express my sincere appreciation to Dr. Donald H. Liles, the supervising professor of my PhD research, for his consistent support, guidance, and inspiration to publish this book. Thanks to Dr. P. Benjamin of Knowledge Based Systems Inc. (KBSI) for his valuable input and time.

I would like to acknowledge the support provided to me by Glide path Inc., Texas, and its president Matt Williams and the Director of engineering Mike Glover. Thanks to Dr. S.N. Imrhan, Dr. Royece Lummus, and Dr. Susan Ferreira for their valuable suggestions.

Finally, I would like to thank the people to whom I owe the most – my family. They have sacrificed just as much, if not more, than I have during my manuscript preparation. Thanks to my wife, Shimu, for her well organized management of our family. Thanks to my little daughters, Mariah and Sarah, for their love & inspiration. They had to be without their daddy many nights while I was busy with working on this book.

September 14, 2007

PREFACE

Recent research has focused on the use of ontologies to promote the sharing of knowledge. Ontologies are becoming increasingly important because they provide the critical semantic foundation for the rapidly expanding field of knowledge. They are very useful for knowledge reuse, knowledge sharing, and enterprise modeling. The use and construction of ontology models have been expanding leaps and bounds among various domains. Modelers from design, manufacturing, medical informatics, logistics, supply chain, systems engineering, etc are building ontology models. They use different tools and methods. Some of these are not well structured and easy to follow. Ontology modeling lacks standard methodology and sufficient literatures. There are very few books available in the market place that talks about ontology modeling methodology and its application and almost none in the area of product and process design.

The objective of this book is to illustrate the ontology modeling methodology and its application in product and process design domain. The ontology modeling methodology discussed in this book is called as the Domain Knowledge Acquisition Process (DKAP) methodology. It is a well structured generic methodology and can be used to model any domain of interest. In this book the DKAP is applied to the product and process design to build a design ontology. In addition, an implementation guideline is also presented to guide the application of the methodology. The implementation guideline provides a step-by-step prescriptive approach to build the ontology. Designing products and processes is very challenging and vital for manufacturing enterprises to stay in the competitive market. Manufacturers want to launch their products in the market as quickly as possible while satisfying their customers in terms of quality, cost and delivery. In order to lunch products early and/or to reduce the product design time, manufacturing enterprises need to have accurate design information readily available at the right time. Design ontology can well serve the above-mentioned purposes.

Design ontology is a hierarchically structured set of terms for describing design domain that can be used as a skeletal foundation for a knowledge base. It can help the collaborative design team by providing accurate design information and guidelines. DKAP is a step-by-step methodology, which captures the product & process design knowledge, stores in reusable format, and shares this knowledge across manufacturing enterprises. DKAP addresses three critical aspect of design ontology. It explores availability of similar domain ontologies for reuse, check accuracy & consistency of captured knowledge, and share captured knowledge.

This book is divided into three major parts. They are part I: ontology modeling, part II: ontology modeling methodology, and part III: a generic product & process design ontology.

Ontology modeling part introduces the reader about available modeling tools and techniques. It describes the traditional as well as future design processes for manufacturing enterprises. Manufacturing enterprises are competing in an environment, which requires the ability to rapidly design their products and processes. This ability requires modeling methods to support design team in providing design information.

In the beginning of the chapter a general overview of enterprise engineering and manufacturing enterprises are described. It also described a detail study of product and process design, which included the importance of product and process design in the manufacturing enterprises, different stages of product design, how typical manufacturing enterprises perform its design functions, and what should be the aspect of future design processes. In the ontology modeling section, objectives and different techniques of ontology modeling are mentioned.

Ontology modeling methodology part presents the detail steps of methodology, including guidelines to build the ontology, and verification process of captured knowledge using consistency matrix.

In the last part generic product design ontology was constructed using the developed methodology. It showed the detail steps of collecting design knowledge, refine them into entities and relations, verify the consistency of information, and publish the ontology. Chapter 5 described each entities of the design domain to the detail including the relations with other entities. Chapter 6 presented an application/demonstration of the method, developed in chapter 4, on two real life companies of building design ontology for their particular product design. The concepts presented in the chapter are equally applicable to other types of manufacturing enterprises including more conventional stand alone enterprises to build domain or site specific ontologies.

Appendix A presents the detail entities and relations of the generic product and process design ontology. Appendix B presents a sample questionnaire to collect domain information and implementation guideline.

WHO SHOULD READ THIS BOOK?

The goal of this book is to explain to more or less anybody involve with the knowledge management. This book presents a methodology and a domain ontology model for generic product and process design which, can serve as a source of design knowledge for enterprise engineering. A direct benefit of such ontology is the enhanced ability to perform product and process design. The ontology, built on IDEF5 language, will allow for common understanding and unambiguous meaning for each design entity and relationship. The result is that different design experts studying the model will have a common basis of communication and understanding. A common understanding is critical in product and process design, especially in collaborative design efforts, which will undoubtedly involve many individuals from different functions, fields and disciplines. This domain ontology can be helpful for both design & research community in following ways:

- Any design team can use this domain ontology for their own product and process design or can get the guideless for their design process.

- Ontology use may reduce the design lead-time & design errors and hence reduce cost of design.

- Facilitates better communication among top management, other departments, suppliers, external partners, sub-contractors, etc.

- Activity model and process model can be derived from an ontology model.

- Ontology developer can use this domain ontology to build similar domain ontology or site-specific ontology in the area of product and process design.

CONTENTS

PART II: ONTOLOGY MODELING METHODOLOGY

PART III: A GENERIC PRODUCT & PROCESS DESIGN ONTOLOGY

APPENDIX

PART I

ONTOLOGY MODELING

CHAPTER 1

INTRODUCTION

1.1 Product and Process Design Overview

Designing product and process is very challenging and vital for manufacturing enterprises to stay in the competitive market. According to current research about 70% of a products cost is determined in the design phase (Mike True & Carmine Izzi, 2001). In some cases of electronics design, design decision can influence up to 80% of total cost. Figure 1.1 represents the influence of major cost function over total cost. Because of high importance of product and process design, manufacturers have great risk if the design process is not appropriate and accurate.

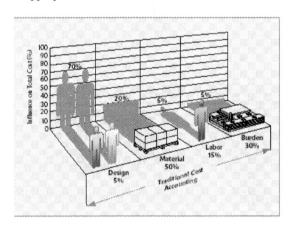

Figure 1.1. The Influence of Design Decision on Total Cost (Mike True & Carmine Izzi, 2001)

Some of these risks include frequent design changes over product development life cycle, budget overrun, schedule overrun, etc. Figure 1.2 shows the design changes over the product development life cycle for traditional design system and integrated product & process design system.

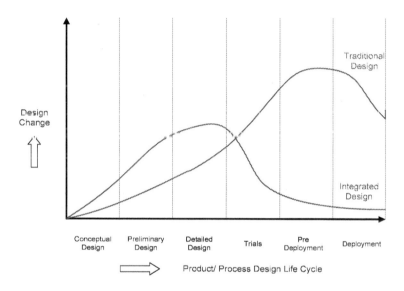

Figure 1.2. Design Changes Over Product Development Life Cycle (Adapted from Kwai-Sang Chin and T. N. Wong, 1999)

Manufacturers want to launch their product in the market as quick as possible. In doing so, they need to expedite their design process. Designers can reduce their design lead-time if they have access to all necessary design information right way. To reduce the cost and product development time, manufacturing enterprises need to share design information and use pre developed proven design templates (Benjamin P., 1995). For this, they have to have comprehensive and integrated design information available.

3

If they have readily available and integrated information, according to research they can reduce their product development time significantly. The following figure 1.3 shows the difference of product development time for integrated product & process design system with the traditional design system.

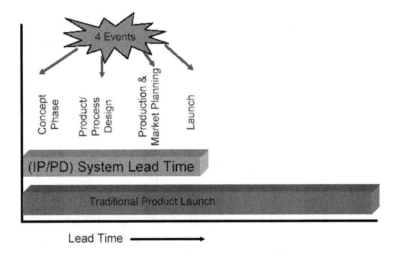

Figure 1.3. Product Development Time for Traditional vs. Integrated Design System
(Martin McGregor, 2005)

Manufacturing Enterprises with integrated design will be seamlessly interconnected among all their internal functions as well as external constituents (William D. Brosey 2001). So that timely, accurate, and consistent design information is available to appropriate design teams. Unfortunately there are lots of enterprises, which are doing everything from the scratch for each of their design needs (A. Gunasekaran, 1998).

To expedite the design process and to survive in the marketplace, manufacturing enterprises have introduced programs of steady improvement of both their products and process design. In doing so they are exploring a variety of concepts, including ERP, Time-based Competition, Rapid prototyping, Quick manufacturing, Continuous Improvement, Process Innovation, and so on. All of these strategies can be successful, but none are panacea. Without accurate & effective design operation of enterprise, no strategies can success for its design needs.

The product realization process is used to transform customer needs into a realized product. According to Dixon and Poli, realization process is a complex set of interrelated activities, both cognitive and physical activities, by which new or modified products are conceived, designed, produced, distributed, serviced, and disposed of (Dixon and Poli, 1995). A customer need for a new or improved product can originate from almost anywhere in the enterprise but majority of the time from sales or marketing groups. Figure 1.4 shows the product realization process with design process in detail. Engineering design is one of the important processes, which translate customer needs into product manufacturing specification that will meet customer requirements (Rudolph, 2005). This book concentrates on how to do engineering design accurately and reduce the product and process design time by sharing design information across manufacturing enterprises.

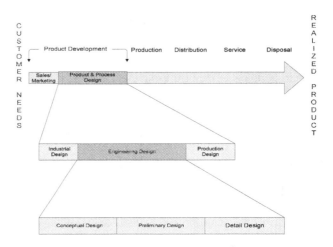

Figure 1.4. Product Realization Process (Modified from Rudolph, 2005)

Product & Process Design is a multi-pronged approach that includes simulation, optimization, collaboration, and by incorporating both existing and emerging knowledge & technologies. Innovation in design will be achieved in the future through the application of these focus areas, while paying particular attention to aspects of concern in design, namely; system performance, cost, and associated manufacturing processes. Product & Process Design consists of three major stages (Tomasz Arciszewski, 2004, Rudolph, 2005, Don Liles, 1995). They are conceptual design, preliminary/ embodiment design and detailed design. Each stage has specific actions and each action has consequences on others followed by that action.

Product & Process Design is an innovative & complex field, and there is no simple 'cookbook' answer for that. Many researchers contributed towards the institutionalization of design process for manufacturing enterprises but still there is no comprehensive design library for manufacturing enterprises (IMTI Inc.2004). This book focuses to achieve the above objective of capturing all design knowledge by developing a methodology using IDEF5.

1.2 Ontology in the Design Process

According to Benjamin P., an important requirement for world-class Product & process design process is the ability to capture knowledge from multiple disciplines and store it in a form that facilitates re-use, sharing, and extendibility. Taxonomies and glossaries, in and of themselves, will not fully address this requirement. There is a perceived need for ontologies rather than mere taxonomies. An ontology is a description of the kinds of things, both physical and conceptual, that make up a given domain, their associated properties, and the relationships that hold among them as represented by the terminology in that domain.

An ontology can be expressed as concepts, relationships and rules about their properties and rules that govern how concepts participate in associations. Ontology Models describe the following:

1) What exists in a domain in terms of objects and events

2) How they relate to one another

3) How they are used inside and outside the boundary of the domain, and

4) Rules that govern their existence and behavior.

The product development lifecycle for most of the product begins with the definition and capture of customer requirements and proceeds through product & process design, manufacturing, and product support. The evolution of a product through these phases involves many transitions of data through the organizations involved with these processes. If sharing information across enterprises are possible, time and costs of product and process design would be reduced. However, because knowledge bases are typically constructed from scratch, each with their own idiosyncratic structure, sharing is difficult. Recent research has focused on the use of ontologies to promote this sharing (Bill Swartout et all, 1996).

One of the most important aspects of the general development and use of an ontology development method is the accumulation of a wide range of domain ontologies. Generally, inefficiency is among the greatest problems in information management. Redundant effort is expended capturing or recreating information that has already been recorded elsewhere (Benjamin P., 1995). Consider an analogy with programming. Different programmers use the same types of routines again and again in different programs frequently. Enormous amounts of time and effort have thus gone into reinventing the wheel again and again. Recognition of this problem has led to the development of vast libraries that have been collected over time that contain often used routines which programmers can simply call straight into their programs, rather than having to duplicate the function of existing code. In product & process design, ontology plays a vital role to share design information among manufacturing enterprises. Figure 1.5 shows the way in which similar kind of manufacturing enterprises can share the relevant information from the same design ontology. In case of sharing appropriate design information, a translator, which could be a person, machine or any other device, is necessary to retrieve the relevant information from shared ontology.

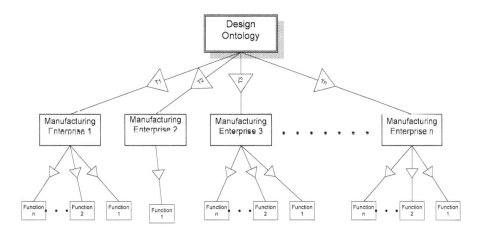

Figure 1.5. Sharing Information among Group of Manufacturing Enterprises

Information management across similar settings faces the same sort of problem (Elisa Finnie, et al, 1997). Manufacturing domains, for example, share many common features; and the more similar the domains, the more features they share. Rather than having to encode this information all over again in every new setting, the idea is to develop an analog of the concept of a programming library by collecting this common information into ontology libraries, i.e., large revisable databases of structured, domain specific ontological information where it can be put to several uses in the envisioned ontology development environment. Two potential uses of such libraries especially stand out (Benjamin P., 1995). First, domain experts developing an ontology for a

9

specific system will be able to import relevant portions of the general database for the kind of system they are describing directly into the ontology knowledge base. This will save them the trouble of having to collect, process, organize, and record the information directly. This information will of course be malleable, so that a given expert can modify it in light of features unique to his or her system. Second, the information can be used to construct general techniques for aiding domain experts in extracting domain knowledge. For example, by isolating and analyzing general patterns or features of ontologies within certain domains, one can develop productive strategies for eliciting and structuring the sorts of knowledge one is likely to find in those domains. For instance, if a certain common kind of machine varies in certain details from location to location, the background ontology database can import the common information directly, and then lead the user through a series of questions to elicit the specifications that are unique to the particular domain. Again, an expert may not know how a certain object should be classified. By searching on a list of essential properties of the object, the tool could return a set of kinds in which the object would most naturally be included.

Different functions within the enterprise can share the operational data from the shared ontology (Robert Jasper and Mike Uschold, 2000). Figure 1.6 shows the architecture of such sharing within the enterprise. Ontology author creates design ontology, which is used by operational data author, who retrieves operational data and put in common database. This database serves relevant functional department by providing relevant data with the help of translators. Translated data must be specified by

the shared ontology to prevent the distortion of information.

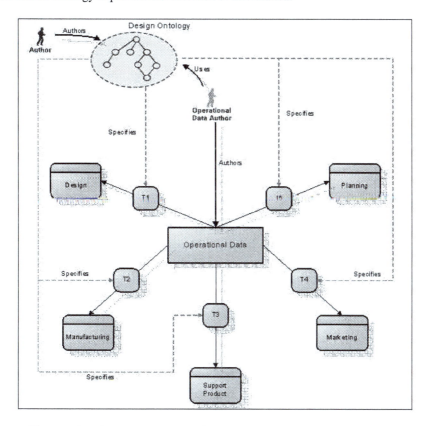

Figure 1.6. Relevant Data Access via Shared Ontology within an Enterprise

1.3 The Need for Design Ontology

Design ontology promotes sharing design information and enables manufacturing enterprises to be competitive in the present marketplace by reducing design lead time and cost of design. Developing an ontology modeling methodology is justified based on the importance of designing product and process in a manufacturing enterprise, the need for a comprehensive method for capturing design knowledge, and the difficulty of developing such a methodology (Shanks et al., 2003). Product and process design plays a very important & critical role for manufacturing enterprises to survive in the future. There are some enterprises that have ignored or unable to incorporate all necessary design information in their design had to exit the market with millions dollars in losses (Pugh, S., 1991).

In recent years a lot of research is going on to develop ontologies. Stanford University Knowledge Systems Laboratory is playing a vital role to develop ontology tools and techniques and sharing related information. Many disciplines now develop standardized ontologies that domain experts can use to share and annotate information in their fields. Medicine, for example, has produced large, standardized, structured vocabularies such as SNOMED (Price and Spackman, 2000) and the semantic network of the Unified Medical Language System (Humphreys and Lindberg, 1993). Broad general-purpose ontologies are emerging as well. For example, the United Nations Development Program and Dun & Bradstreet combined their efforts to develop the UNSPSC ontology, which provides terminology for products and services. There is some research done on ontology modeling for some product attributes not on the product design. Some of them are cost ontology for enterprise modeling (Tham, K.D., 1999), ontology of quality for enterprise modeling (Kim, H.M. et al., 1999), ontology for modeling and analyzing of enterprise competence (Gruninger, M., et al., 2000), etc. In the mid 90's European researchers tried to develop design ontology named 'spirit'

12

but did not go through (Benjamin P., 2005). Benjamin P. of KBSI Inc. developed an ontology for acquiring CIM for manufacturing enterprise in the late 90's (Benjamin P., 1995). After that there are no research found to develop such ontology for product & process design for manufacturing enterprises.

CHAPTER 2

DESIGNING PRODUCTS AND PROCESSES

2.1 Introduction

This chapter discusses a general description of enterprise engineering, manufacturing enterprises, and product and process design in details. In the product and process design, its importance within the manufacturing enterprises, how typical enterprises perform their design task, and stages of design process are discussed. In the later section, discussions are included of ontology modeling, available techniques of modeling, different tools used in ontology modeling, benefits of such modeling, and details of IDEF5 modeling tool.

2.2 Enterprise Engineering

Enterprise Engineering consist of people, process and technology and is defined as that body of knowledge, principles, and practices having to do with the analysis, design, implementation and operation of an enterprise (Don Liles et al., 1996). It is a set of activities organized into business processes cooperation to produce a set of desired results (Presley, 1997). It helps to understand the mechanism behind the operation of enterprise. It is a complex system and without understanding the nature of the complexity, it is quite impossible to run the enterprise efficiently.

An enterprise model is a computational representation of the structure, activities, processes, information, people, behavior, goals and constraints of a business, government or other enterprise. It can be both descriptive and definitional—spanning what is and what should be. The role of an enterprise model is to achieve model-driven enterprise design, analysis, and evaluation. (Michael G. et al., 2000). However, many approaches to enterprise modeling lack an adequate specification of the semantics of the terminology of the underlying enterprise models, which leads to inconsistent interpretations and uses of knowledge. Ontology modeling is one of the enterprise models, which can incorporate semantics of the terminology into the ontology.

2.3 Manufacturing Enterprises

Manufacturing enterprises need to perform a variety of functions to do business. Some of these functions are very important compared to others and the degree of importance varies from enterprise to enterprise. All manufacturing enterprises may not have to perform all functions, it depends on the nature of their business. Typically, a simple manufacturing enterprise need following six functions for their business (Don Liles, 1995). Figure 2.1 shows these six functions with their relations, inputs, outputs, and mechanism. A brief description of these functions is as follows (MD Sarder and Don Liles, 2005).

Strategic planning- is an overall enterprise directive for the development of tactical level plans, policies, and procedures. It specifies the role of the organization with respect to the environment and defines the enterprise culture. It consists of following sub functions.

15

- Defining business

- Setting required performance

- Creating, evaluating, and selecting strategies

Resource management- is concerned with the management of resources including capital, personnel, information, and facilities. Manage resources is a tactical level activity where the strategic level objectives regarding resources are translated into tactical plans which define expected contributions of each operational area toward achieving the strategic plan. This includes following activities.

- Managing finance / accounting

- Managing personnel

- Managing facilities

- Developing aggregate plan

- Managing information system

Marketing- provides the enterprise with a dynamic external link to its customers and the environment. Marketing information gathered during analysis serves as a basis for developing forecasts, strategic plans, tactical plans, and the development of advertising. Sub functions of marketing are as follows.

- Analyzing marketing information

- Developing plans & rules

- Marketing product

- Selling product

Design- translates customer needs into design directives/ specifications. This is one of the core functions for manufacturing enterprises. Design consists of following activities.

- Supervising engineering functions

- Developing conceptual design

- Developing embodiment design

- Developing detailed design

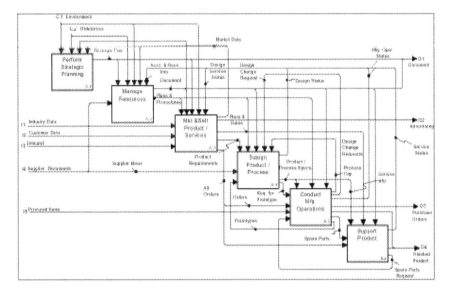

Figure 2.1. IDEF0 Representation of Manufacturing Functions (Sarder & Liles, 2005)

Manufacturing- produces realized products according to design directives. This means ensuring that adequate resources (material, information, equipment, and labor) are available to meet production requirements, and providing a release mechanism, which controls the loading of the shop floor. This includes following activities.

- Planning manufacturing

- Supervising shop floor activity

- Executing & controlling manufacturing

- Supporting production

- Maintaining facilities

- Distributing product

Support service- after a product is manufactured and sold, certain support functions might be necessary. These might include issuing manuals and documentation, providing training and logistics support, providing repair services, or issuing spare parts. It includes following activities.

- Supervising product support

- Producing documentation

- Providing training

- Providing repair service

- Providing spare parts

Manufacturing enterprises contribute most of nation's GDP. According to recent statistics, U.S. manufacturing, if it were a nation, would be the world's eighth-largest economy. Manufacturing accounts for three-fourths of U.S. exports and the majority of private sector R&D. It also accounts for 13 percent of U.S. GDP and 11 percent of employment. The future prospects for the U.S. economy depend heavily on the vitality of U.S. manufacturing.

World-class manufacturers have established as an operating goal that they will be world class. They assess their performance by benchmarking themselves against their competition and against other world-class operational functions, even in other industries.

They use this information to establish organizational goals and objectives, which they communicate to all members of the enterprise, and they continuously measure and assess the performance of the system against these objectives and regularly assess the appropriateness of the objectives to attaining world-class status.

World class manufacturers integrate all elements of the manufacturing system to satisfy the needs and wants of its customers in a timely and effective manner. They eliminate organizational barriers to permit improved communication and to provide high quality products and services.

2.4 Product & Process Design

It is a process of transforming customer needs into realized product through complex set of design activities (Rudolph, 2005). A detail discussion of product and process design is presented in the following sections.

2.4.1 Importance of Product & Process Design within Manufacturing Enterprise

As mentioned earlier that Product and process design plays a very important & critical role for manufacturing enterprises to survive in the future. Among all functions of manufacturing enterprise, product and process design is the single most important function. According to current research about 70% of a products cost is determined in the design phase. In some cases of electronics design, design decision can influence up to 80% of total cost.

2.4.2 Stages of Product & Process Design

Though product and process design is an innovative process, it follows structured steps of design activities. A product design evolves over time in design phases from the identification of customer needs to the realized product. This design stages are time sequential in nature. For example, without identifying customer needs it is not possible to do the detail design. Though concurrent design is necessary to reduce the design time, there is a limit for the extent of concurrency. There are many authors, who have mentioned the different stages of design process in their literature (John Priest and Jose Sanchez, 2001, Pahl and Beitz, 1996, Dixon and Poli, 1995, Rudolph, 2005, etc). The majority of researcher proposed three major stages of design processes. They are conceptual design, embodiment design, and detail design. In the conceptual design, designers synthesize a variety of candidate working principles or concepts. In the embodiment design, designer conduct lots of trade of analysis, determine values for the design parameters, and select the best candidates. Detail design phase produces the detail specification of engineering design.

2.4.3 Traditional Ways of Doing Product & Process Design

Traditional design process is iterative in nature and has very less integration of design information. Design team starts with concept generation and follows consequent design steps. The changes of design along product life cycle for traditional design are shown in the following figure 2.2. The more the number of changes, the more the cost

of design is. The number of changes reach maximum during pre-deployment.

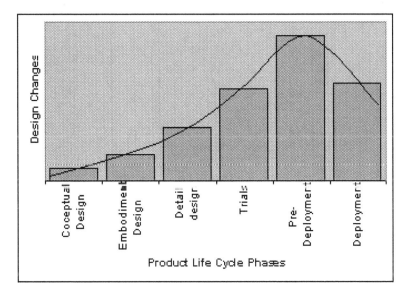

Figure 2.2. Design Changes across Product Life Cycle

Again the more changes in the later stages of design cycle, the more the cost to

fix those problems are. Figure 2.3 shows the relation of correction costs and product life

cycle. The exponential nature of cost correlation sometime makes impossible to fix the

design problems.

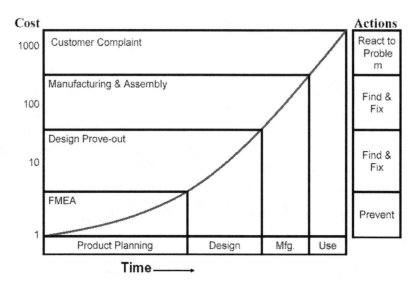

Figure 2.3. Cost of Correction over Time (Martin McGregor, 2005)

2.4.4 Product & Process Design in the Future

Product and process design have a major influence on the competitiveness of the enterprise, hence it is especially critical that the design information must be accurate and design function be better integrated with the other functions of the enterprise. The future design will bring more mature product designs, which can be more effectively produced within a company's existing or planned production system and more effectively supported. New product design and introduction lead-time or time-to-market will be reduced to meet rapidly changing technology and customer demands and increase enterprise flexibility. According to Kenneth Crow, the objectives of future product and process are:

- The design of products to better meet customer needs and quality expectations

- The design of processes or the consideration of process capabilities in designing products in order to produce products at a more competitive price

- Reduction of product and process design cycle time or time-to-market to bring products to market earlier

- High productivity through release of producible designs and minimization of disruptive design changes

To accomplish these objectives, following approaches are necessary for manufacturing enterprises.

- Alignment of product development with business strategy;

- Integrated and collaborative approach to product and process design;

- Extensive reuse of design knowledge (Design Ontology);

- Wide access of accurate design information;

- Optimization of the product and process design to enhance manufacturability, testability, affordability, reliability, maintainability, etc

The re use of design information will significantly reduce the design time. At the same time design team will capture knowledge for future use from different sources such as new finding from its own design process and individual expertise from its own human recourses, which includes sales and marketing people, design people, manufacturing stuffs and so on.

CHAPTER 3

ONTOLOGY MODELING: TOOLS & TECHNIQUES

3.1 What's in the Ontology Model?

In an ontology model, descriptors are cataloged (like a data dictionary) and create a model of the domain, if described with those descriptors. Thus, in building an ontology, it must produce three products, which are to catalog the terms, capture the constraints that govern how those terms can be used to make descriptive statements about the domain, and then build a model that when provided with a specific descriptive statement, can generate the "appropriate" additional descriptive statements.

In the context of knowledge sharing, the term ontology means a specification of a conceptualization. That is, an ontology is a description (like a formal specification of a program) of the concepts and relationships that can exist for an agent or a community of agents. This definition is consistent with the usage of ontology as set-of-concept-definitions, but more general. And it is certainly a different sense of the word than its use in philosophy. Ontologies are often equated with taxonomic hierarchies of classes, class definitions and the subsumption relation, but ontologies need not be limited to these forms. Ontologies are also not limited to conservative definitions, that is, definitions in the traditional logic sense that only introduce terminology and do not add any knowledge about the world. To specify a conceptualization, one needs to state axioms that do constrain the possible interpretations for the defined terms.

Any domain with a determinate subject matter has its own terminology, a distinctive vocabulary that is used to talk about the characteristic objects and processes

24

that comprise the domain. The nature of a given domain is thus revealed in the language used to talk about it. Clearly, however, the nature of a domain is not revealed in its corresponding vocabulary alone; in addition, one must (i) provide rigorous definitions of the grammar governing the way terms in the vocabulary can be combined to form statements and (ii) clarify the logical connections between such statements. Only when this additional information is available is it possible to understand both the natures of the individuals that exist in the domain and the critical relations they bear to one another. An ontology is a structured representation of this information. More exactly, an ontology is a domain vocabulary together with a set of precise definitions, or axioms, that constrain the meanings of the terms in that vocabulary sufficiently to enable consistent interpretation of statements that use that vocabulary.

3.2 Objectives of Ontology Modeling

The primary goal of the ontology modeling is to provide a structured technique, supported by automated tools, by which a domain expert can effectively develop and maintain usable, accurate, domain ontologies.

A key to effective integration is the accessibility of rich ontologies characterizing each of the domains addressed by each cluster. For instance, access to a manufacturing ontology that includes constraints on how a given part is manufactured can aid designers in their design of a complex product by giving them insight into the manufacturing implications of their design concepts. Similarly, access to an engineering ontology that includes constraints on how a given part is to function given a particular shape or fit can aid process planners in their development of the appropriate manufacturing processes. A commonly accessible collection of relevant ontologies thus permits more efficient sharing of information arising from various sources within the enterprise.

An enormous problem in the coordination of collaborative team is the diversity of backgrounds the various kinds of team members bring to their respective roles. As a consequence, many members use similar terminology in many different ways with many different connotations. Because of such differences, the information that one member intends to convey to another may in fact become garbled; in the best case, such miscommunications can be responsible for a great deal of lost time and resources. Consequently, it is often necessary in the course of a large project to standardize the relevant vocabulary. The ontology capture method provides a principled method for carrying out this task efficiently and effectively, and maintaining the results of the task in a robust, accessible form.

3.3 Modeling Techniques

Over the last few decades numerous conceptual modeling techniques, used to define requirements for building information systems, processes, activities, etc. have emerged with no consistent theoretical foundation underlying their conception or development. Concerned that this situation would result in the development of models that were unable to completely capture important aspects of the real world, WAND and WEBER (Wand and Weber, 1989; Wand and Weber, 1990; Wand and Weber, 1993; Wand and Weber, 1995) developed and refined a set of models based on an ontology defined by BUNGE (Bunge, 1977) for the evaluation of modeling techniques. These models are referred to as the BUNGE-WAND-WEBER (BWW) models, or the BWW ontology. Ontology studies the nature of the world and attempts to organize and describe what exists in reality, in terms of the properties of, the structure of, and the interactions between real-world things (Shanks et al., 2003). As computerized information systems are representations of real world systems, WAND and WEBER suggest that ontology can be used to help define and build information systems that contain the necessary representations of real world constructs. The BWW representation model is one of three theoretical models that make up the BWW models. Its application to information systems foundations has been referred to by many researchers (for an overview see (Green and Rosemann, 2004)) and is now often referred to as simply the BWW model. Some minor alterations have been carried out over the years (Wand and Weber, 1993; Wand and Weber, 1995; Weber, 1997), but the current key constructs of the BWW model remains same.

IDEF models are widely used in the mid 90's for various modeling such as activity modeling, process modeling, information modeling, ontology modeling, etc. Among all of these modeling, activity modeling and process modeling are very familiar. In the past few years there were some Object Oriented modeling techniques such as UML and Protégé used in ontology modeling.

3.4 Tools Used in Ontology Modeling

There is handful of tools for ontology modeling. The first tool used in ontology modeling is Petri net in 1962 and recent developed tools include BPMN, UML, Protégé, etc. As Mr. John Zachman in his seminal work on information systems architecture observed, "... there is not an architecture, but a set of architectural representations. One is not right and another wrong. The architectures are different. They are additive, complementary (Zachman 87).

IDEF suites were developed in the mid 90,s and are proven useful. Thus, IDEFØ provides a compact, yet surprisingly powerful, conceptual universe for modeling business activities; for all its power, however, it would be highly inconvenient, if possible at all, to use it to design a relational database; IDEF1X is the method that is optimized for that task. Similarly, IDEFØ explicitly excludes temporal information, and limits what can be represented about temporal relations that hold between business activities, as well as the objects involved in the internal structure of those activities. These exclusions are what give IDEFØ its power in modeling business activities. For in a method design as in a programming language design, what distinguishes a well designed effective method is what is left out more so than what is left in. IDEF3, on the other hand, includes explicit representations of processes, time intervals, and temporal relations and, hence, is ideally suited for expressing information about timing and sequencing; it also includes the capacity to express arbitrary information about the individuals participating in those processes. It lacks, however, the specialized representations of IDEFØ and, therefore, information that IDEFØ expresses with great ease and simplicity is, by comparison, expressed only awkwardly in IDEF3. The connection between these methods and IDEF5 is rather straightforward. Of the methods just mentioned, the IDEF5 schematic language is perhaps closest to IDEF1 and

IDEF1X. However, the connection between IDEF1/1X and IDEF5 is analogous to that between IDEFØ and IDEF3.

The information in an IDEF1 or IDEF1X model could in principle be expressed in the IDEF5 elaboration language. However, because it does not contain the well-designed, specialized representations of IDEF1/1X, it would be exceedingly cumbersome in IDEF5 to design a relational database, for example. But the expressive power of IDEF1/1X soon reaches its limits and, hence, could not possibly do all that is expected of a general ontology language.

In a sense, the designs of both IDEF3 and IDEF5 break the traditional mold according to which methods are purposely designed with limited expressive power. The elaboration languages of both methods are full first-order languages (and more besides) and, hence, are capable of expressing most any information that might need to be recorded in a given domain. This break with tradition not only reflects the need for greater expressive power, but also reflects the development and increased utilization of more intelligent tools and automated, model-driven systems in business and engineering. Intelligent tools and model-driven systems generally must manipulate much richer forms of information than can be expressed in a traditional method. This motivates the design of richer methods that have the capacity to represent and organize such information, methods that are not restricted to pencil and paper form and, hence, which truly augment the ability of human agents to create, manage, and reuse a richer store of knowledge.

IDEF5 is being developed in the belief that it can contribute in a vital way to the realization of this vision of global knowledge sharing. The IDEF5 method therefore fulfills an important need by providing a cost-effective mechanism to acquire, store, and maintain scaleable and re-usable ontologies. The intended contribution of IDEF5 is a method to guide and assist domain experts and knowledge engineers in the construction of both small and large reusable ontologies. Figure 2.4 shows the historical uses of different ontology tool to build ontologies.

The BWW representation model has been used in over twenty-five research projects for the evaluation of different modeling techniques (see (Green et al., 2005) for an overview). In this section, we briefly summarize those studies that focus on process modeling techniques. KEEN and LAKOS (Keen and Lakos, 1996) determined essential features for a process modeling scheme by evaluating six process modeling techniques, among them ANSI flowcharts, Data Flow Diagrams (DFD), in a historical sequence by using the BWW representation model.

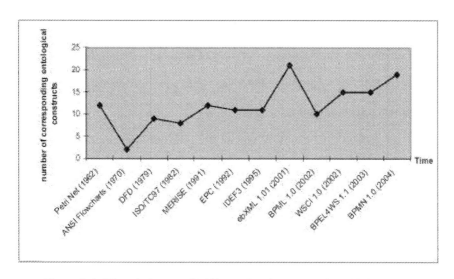

Figure 2.4. Historical Uses of Different Ontology Tools in Building Ontology

GREEN and ROSEMANN (Green and Rosemann, 2000) analyzed the EPC notation with the help of the BWW ontology, focusing on both ontological completeness and clarity. Their findings have been empirically validated through interviews and surveys (Green and Rosemann, 2002). Confirmed shortcomings were found in the EPC notation with regard to the representation of real world objects and business rules, and in the thorough demarcation of the analyzed system.

GREEN et al. (Green et al., 2005; Green et al., 2004) compared different modeling standards for enterprise system interoperability, including Business Process Execution Language for Web Services v1.1 (BPEL4WS), Business Process Modeling Language v1.0 (BPML), Web Service Choreography Interface v1.0 (WSCI), and ebXML Business Process Specification Schema (ebXML BPSS) version 1.1. All these standards, which proclaim to allow for specification of intra- and inter-organizational

31

business processes, have been analyzed in terms of their ontological completeness. The study found that ebXML provides a wider range of language constructs for specification requirements, with this situation being clearly indicated through its relatively high degree of ontological completeness.

There has been further work concentrating on the ontological evaluation of process modeling related techniques, see, for example, (Opdahl and Henderson-Sellers, 1999; Soffer et al., 2001) Some of these techniques rely on an object-oriented (OO) paradigm (like UML, OML, OPM). Stanford University research center developed well-known Protégé tool for ontology modeling. This tool is being widely used to build ontology because of its automated sharing information feature.

As mentioned earlier that IDEF5 is a powerful tool to capture design knowledge and build the design ontology. It is easy to use and step-by-step method. In the IDEF5 method, capturing the content of certain assertions about real-world objects, their properties, and their interrelationships and representing that content in an intuitive and natural form construct ontology. An ontology characterizes what exists: the kinds, their properties, and their interrelationships in a given domain, as revealed in the terminology used by experts in the domain. A complete ontology, then, reveals the fundamental nature of a given domain. The construction of an ontology differs from traditional information capture activities in the depth and breadth of the information captured. Thus, an ontology development exercise will go beyond asserting the mere existence of relations in a domain. From the above discussion it has been shown that IDEF5 is a powerful tool to build ontology but never got exposure. It is better than Protégé in terms of graphical presentation as well as comprehensiveness.

3.5 Benefits of Ontology Modeling

Ontologies are being constructed for a growing number of manufacturing, engineering, and scientific domains. With such ontologies in place, many benefits could be realized on a global scale: standardized terminology with precise meanings that are fixed across industries and across international borders, and the ability to access and reuse a huge number of ontologies in the design and construction of new systems. Central products of this effort include the Knowledge Interchange Format (KIF), a text-based logical language for the interchange of knowledge, and Ontolingua, a mechanisms built on KIF for translating knowledge between different representation languages. The UML, Protégé, and IDEF5 methods have been designed with the Knowledge Sharing Effort and its vision closely in mind. Most notably, the IDEF5 elaboration language is the central medium for storing ontology information collected via the IDEF5 method - uses KIF as its foundation, and is thus wholly compatible with the central tools of the Knowledge Sharing Effort. This is particularly crucial as the concepts behind the effort become even more widely accepted and implemented.

Ontology development provides several benefits to organized enterprises. The benefits of ontology development can be grouped under two headings:

1. Benefits of developing the ontology: The process of ontological analysis is a discovery process that leads to an enhanced understanding of a domain. The insights of ontological analysis are useful for (i) identification of problems (diagnosis), (ii)

identification of the problem causes (causal analysis), (iii) identification of alternative solutions (discovery and design), (iv) consensus and team building, and (v) knowledge sharing and reuse.

2. Benefits derived from the products of ontology development: The ontologies that result at the end of an ontology development effort can be used beneficially for (i) information systems development: ontologies provide a blueprint for developing more intelligent and integrated information systems, (ii) systems development: ontologies can be used as reference models for planning, coordinating, and controlling complex product/process development activities, (iii) business process reengineering: ontologies provide clues to identifying focus areas for organizational restructuring and suggest potential high-impact transition paths for restructuring.

Ontological analysis and development have been shown to be useful for: (i) Consensus building, (ii) Object-oriented design and programming, (iii) Component based programming, (iv) User interface design, (v) Enterprise information modeling, (vi) Business process reengineering, and (vii) Conceptual schema design.

PART II
ONTOLOGY MODELING METHODOLOGY

CHAPTER 4

DEVELOPMENT OF THE DKAP METHODOLOGY

4.1 Introduction

In developing design ontology, a well-structured methodology is essential to capture appropriate design knowledge. This methodology is for creating an ontology of product and process design using IDEF5 for manufacturing enterprises, which enable them to be competitive in the present marketplace by reducing design lead time and cost of design. This step by step methodology is a systematic engineering approach to capture the knowledge of product & process design domain, represent the relationship, and share this knowledge across manufacturing enterprises and make it efficient to operate. This methodology will help the manufacturing enterprise to

- Capture the knowledge of their interest

- Share common understanding of the structure of information across the organization

- Enable reuse of captured knowledge & domain knowledge

- Make explicit domain assumptions

- Separate domain knowledge from the operational knowledge

- Analyze domain knowledge

A cross consistency matrix was developed to measure the level of consistency and accuracy of information captured and represented by the ontology model. A good reusable ontology must concern with the consistency and accuracy of communication between organizations (Yeh, I. *et al,* 2003), systems, and misinterpretations in communications are addressed by ontologies that explain and reconcile terminology, jargon, and nomenclature specific to each party.

4.2 Knowledge Capturing Method

An important requirement for world-class Product & process design process is the ability to capture knowledge from multiple disciplines (P. Benjamin, 1995) and store it in a form that facilitates re-use, sharing, and extendibility. As companies face increasing competition in global markets, there is a renewed understanding that operational expertise and crisp execution provide significant business advantages. The ability for an enterprise to capture and share appropriate knowledge within the enterprise will translate into both operational and financial benefits (Snow John, 2004).

Knowledge captures and knowledge sharing in the area of product/ process design has the following benefits:

- Reduced product development time
- Decreased duplication of effort, increased collaboration and reduced design times
- Maximized productivity and flexibility of the design team
- Better knowledge transfer among different functional departments
- Maximized resource utilization

The knowledge capture issue is often discussed in terms of capturing explicit and tacit knowledge. Explicit knowledge is that which can be expressed in language and can therefore be codified and recorded (Merali & Davis, 2001). Tacit knowledge is that which cannot be expressed in language. It is generally accepted that tacit knowledge can be transmitted through socialization processes (Nonaka & Takeuchi, 1995) such as a master apprentice "learning by accompanying, watching, helping and copying" arrangements. Most organizational action is context-specific, and tacit knowledge underpins the choice of appropriate actions for given situations. The followings are some of the ways, which could be used to capture this knowledge.

- Written documentation

- Case based reasoning

- Ontology building

Among all of them, ontology is the latest form of knowledge capturing method; where information with specific context can be stored. It is visible, easy to understand and easy to share and reuse. One of the most important aspects of the general development and use of an ontology acquisition method is the accumulation of a wide range of domain ontologies. Generally, inefficiency is among the greatest problems in information management. Redundant effort is expended capturing or recreating information that has already been recorded elsewhere (Benjamin P. et al, 1995). Consider an analogy with programming. Different programmers use the same types of routines again and again in different programs frequently. Enormous amounts of time and effort have thus gone into reinventing the wheel again and again. Recognition of this problem has led to the development of vast libraries that have been collected over time that contain often used routines which programmers can simply call straight into their programs, rather than having to duplicate the function of existing code.

4.3 Nature of Ontology

Ontology development methods can effectively capture knowledge about objects in the real world and the associations that exist between people, places, machines, events, etc. Ontologies provide the background context within which information is transferred between two agents.

In the context of product and process design, an ontology is a computationally tractable representation of what kinds of things and associations experts recognize in a given domain. An ontology includes much more that just "dictionary" information. In addition to the dictionary sort of information, an ontology also characterizes the acceptable inferences that a domain expert can make if he or she is given a statement made from terms in that dictionary. An ontology seeks to identify the primary classes, or kinds, of objects that are within the domain by isolating the properties that define the members of those kinds and the characteristic relations that hold between domain objects. Such representations are purposely structured in a way that closely reflects human conceptualization of the domains in question (Richard J. Mayer *et al.*, 1993). Thus, differing perspectives on the same domain and their interrelations are also supported. This feature significantly distinguishes a full-blown ontology from a traditional knowledge base.

4.4 Characteristics of Design Ontology

In modern design environments products are so complex that correct externalized information and knowledge must be readily accessible by the designer. Unlike any other function in the manufacturing enterprise, design is the most important in terms of cost allocation. As mentioned earlier that about 70% to 80% of a products cost is determined in the design phase. In that case manufacturers have a great risk if the design process is not appropriate. Manufacturing enterprises cannot afford to have a design on wrong information. It is widely believed that design engineers spend up to 47% of their time seeking design information in the design process (Hales, 1987). Given that design can be described as a problem solving process and considering that engineers tend to solve problems based on available knowledge. It is important to ensure that appropriate knowledge is available at the correct time in the process (Lawson 1990; Cross 1994; Hubka 1996; Pahl and Beitz 1996). Knowledge can be made available in two ways, i.e. external (established from the experience of others, existing research and new research) or internal (designers own experience and knowledge established through learning) (Lawson 1990). Both external and internal knowledge can be represented by the design ontology.

An appropriate design representation scheme is a prerequisite for an efficient design system. As the design ontology aims to support the evolutionary process of design reuse, and its representation scheme should be able to capture the information and knowledge involved at all stages of the design process. Much of the researches have been devoted to issues relating to problem clarification, conceptual design, and detailed design. Less effort has been made to the embodiment design representation, especially from a computational perspective (Ong & Guo, 2004).

Design ontology could be any of the three levels of ontologies have been presented as shown in Figure 4.1. These levels are useful when conducting an ontology-building effort. The first is the site-specific ontology. This describes all of the relevant concepts, terminology, structures, and relationships for a specific industrial site. For example, GM plant in Arlington might create ontology to describe its facilities. The second type of ontology is known as practice ontology. Practice ontologies are models of an entire industry. For example, a group of automobile manufacturing companies might develop an ontology for the automobile industry as defined by the companies. The third type of ontology is the domain ontology. This represents all of the information known about a general domain. For example, one might develop a domain ontology for automobile manufacturing in general that includes new research from universities that has not yet been incorporated by industry.

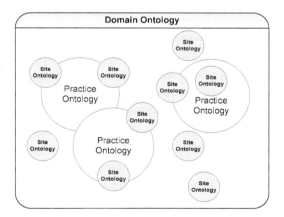

Figure 4.1. Levels of Ontologies (Modified from Benjamin P., 1995)

As Figure 4.1 shows, there is often overlap among the levels of ontologies, and all site specific and practice ontologies are subsets of the domain ontology. Generally the design ontology is a domain specific ontology and can be converted to other according to need. The design ontology must posses some/ all of the following characteristics;

- Accurate, consistent, and up-to-date information

- Easily sharable & reusable

- Enable to capture future research and lesions learned

- Visual & structured representation

4.5 A Brief Description of the Nature of Methodology

This section briefly describes method for capturing product/ process design ontologies. The description of the method focused on the ontology capture procedure as well as the illustrations in using this procedure along with a graphical language. Illustrations drawn from the product and process domain was used to demonstrate the utility of the method.

It is necessary to remember that ontology is a model of reality of the world and the concepts in the ontology must reflect this reality. And it is also necessary to remember that there is no single correct way to do that. Some fundamental rules in ontology design are as follows (Natalya & Deborah, 2001);

- There is no one correct way to model a domain— there are always viable alternatives. The best solution almost always depends on the application that designer have in mind and the extensions that designer anticipate.

- Ontology development is necessarily an iterative process.

- Concepts in the ontology should be close to objects (physical or logical) and relationships in design domain of interest. These are most likely to be nouns (objects) or verbs (relationships) in sentences that describe design domain.

As noted earlier, ontology is a documentation of the terminology used to describe objects, properties, and associations in a particular domain. It also includes the rules for combining and using that terminology to form statements about the domain, and identifies the sanctioned inferences that can be made from those statements in the domain. This use of "ontology" is consistent with the traditional use because what "exists" in a given domain is largely influenced by the ability of the agents to individuate or "carve up" the world. Because humans differ greatly in this ability due to differing capacities and differing conceptual viewpoints, ontologies are rarely perspective-invariant. Ontology development–the analysis and detailed characterization of the terminology in a domain–is focused on understanding the concepts of a domain from these varied perspectives. It is also focused on extracting the essential nature of these concepts and representing this knowledge in a structured manner (Benjamin P., 1995).

Ontology in the product and process design plays an important role in the future product and process design for manufacturing enterprises. Traditional design process translates customer needs into design specifications, in doing so, most cases it does not use any pre-developed design template or design ontology (Dixon & Poli, 1995, Rudolph, 2005). Neither it capture its own design knowledge for future use. Figure 4.2 shows input, output and constraints of traditional design process.

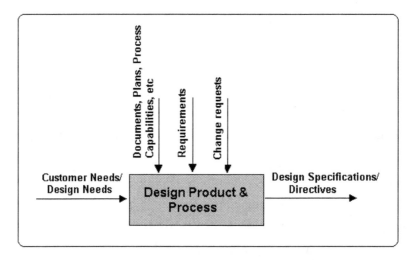

Figure 4.2. Traditional Product & Process Design

The future design system will work as like usual except the design process will re-use ontology, already develop by its own or by others. This re-use of design information will significantly reduce design time. At the same time this system will capture knowledge from different sources such as new finding from its own design process and individual expertise from its own human recourses, which includes sales and marketing people, design people, manufacturing stuffs, and so on. Other design information resources include design experts from outside the enterprise, available resources such as books, journal articles, seminars, etc, lesson learned from different enterprises, and emerging knowledge and techniques such as university research or new software/ tool which is not published yet. Once this information is captured and builds the design ontology, it needs to publish the ontology for others to reuse. Figure 4.3 describes the new architecture of future design process compared to traditional design process. To meet the future needs of he design processes, the methodology must have the provision to re-use already developed design ontology as is or with minor change. Another significance of this methodology is to check the consistency and accuracy of captured information, which is vital for design ontology. It also publishes the developed design ontology into shared ontology repository for others to re-use.

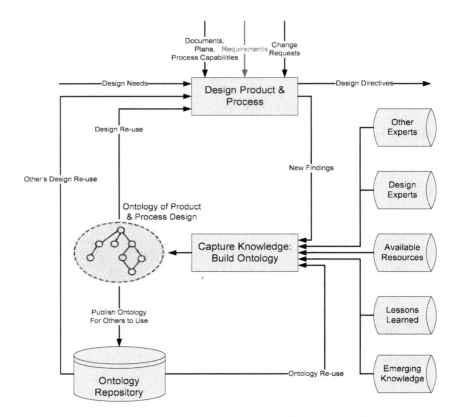

Figure 4.3. Future Product & Process Design for Manufacturing Enterprise

Design process varies from industry to industry. For instance, electronic industry's design process differs from that of automobile industry because of different product requirements. Similarly automobile industry's design process will be different from the design of aircraft industry. But overall design process more or less similar to all industries. Existing literature has anecdotal stories of enterprises doing their product and process design. By studying these design process, current practices of product and process design was identified. This process serves as a guideline for developing design ontology. As mentioned earlier, the design process must be accurate and industry specific.

Ontology modeling is comparatively a new but expanding research area. So there is no common technique available to build the ontology. Current literature describes the tools and techniques that have been successfully deployed by various authors/ enterprises to build their ontology. Ontology authors use different techniques and tools to build their ontology according to their specific needs. Some techniques and tools are more useful over others.

4.6 Structure of the DKAP Methodology

In developing design ontology, a well-structured methodology is essential to capture appropriate design knowledge. This Domain Knowledge Acquisition Process (DKAP) methodology is constructed to meet the particular design needs. Different ontology author uses different steps to build their ontology. For example Ricardo Calmeta mentioned four steps, Natalya & Deborah McGuinness mentioned seven steps and P. Benjamin mentioned five steps to build their own ontologies. All of them mentioned about the following steps.

- Organize and Scope the Project
- Collect Data
- Analyze Data
- Develop Initial Ontology
- Refine and Validate Ontology

All of the above-mentioned steps do not completely meet the need for the design ontology. The steps of the Design Knowledge Acquisition Process (DKAP) methodology of product and process design are slightly different from other ontology building approach to tailor the capture of design knowledge. It has nine major steps. They are as follows.

1. Determine the domain & Scope of the ontology

2. Check availability of existing ontologies

3. Organize the project

4. Collect and Analyze Data

5. Develop Initial Ontology

6. Refine and Validate Ontology

7. Check consistency & accuracy of ontology

8. Collect additional data and analyze data

9. Incorporate lessons learned and publish ontology

The structure of this DKAP methodology for ontology development process is shown in figure 4.4. Two fundamental differences between this particular design ontology and an ordinary ontology are as follows. DKAP is capable to re-use already developed design ontology as is or with minor change.

Another significance of this methodology is to check the consistency and accuracy of captured information, which is vital for design ontology. It also publishes the developed design ontology into shared ontology repository for others to re-use. As mentioned earlier, design process is almost similar for different products but very

48

important for each products and processes. A common repository of design ontologies

is very useful for many manufacturing enterprises for their design needs.

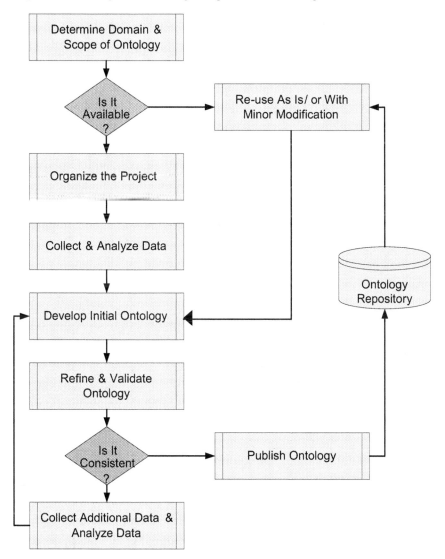

Figure 4.4. Structure of DKAP Methodology

4.7 Detail Steps of the DKAP Methodology

The detail steps of the methodology are discussed in the following sections along with an example case.

Step 1. Determine the domain & Scope the ontology: This activity will establish the purpose, viewpoint, and context for the ontology development project and assign roles to the team members. The purpose statement provides a "completion criteria" for the ontology description capture effort. The purpose is usually established by a list of 1) statements of objectives for the effort, 2) statements of needs that the description must satisfy, and 3) questions or findings that need to be answered. For example, the purpose statement for this situation is: *"To develop a design ontology of a generic product and process design".*

Once the purpose of the effort has been characterized, it is possible to define the context of the project in terms of 1) the scope of coverage, and 2) the level of detail for the ontology development effort. The scope defines the boundaries of the description development effort, and specifies which parts of the systems need to be included and which are to be excluded.

Establishing viewpoints is important to develop the ontology. It is related to the purpose of development. For instance, collaborative design team will normally use design ontology, hence it is appropriate to establish viewpoints with respect to collaborative design team. To say "an ontology is in the eye of the beholder" may be too extreme a view. Nevertheless, the role of differing viewpoints on the outcome of ontology capture efforts is an important one. The differences in viewpoints are often reflected in different aspects of the ontology such as the specification of the level of detail of the description capture. Table 4.1 shows an IDEF 5 form of ontology description summary including purpose, context and viewpoints.

Table 4.1. Definition of the Ontology Development Project

Ontology Description Summary Form			
Project: Automobile Design Ontology	Analyst: Md Sarder	Reviewer: Don Liles	Document Number:
Version: 1	Date: 9/12/2005	Date: X/X/2005	
Purpose: To develop an ontology of the Product & Process Design domain for Automobile manufacturing enterprises. The resulting description must serve 1) as a knowledge repository for Company A's design system integration project and 2) as a reference model for Automobile industry as a whole.			
Context: The information acquired must be sufficient to organize design activities, specify precedence relationships, and supports world-class design procedures.			

Step 2. Check availability of existing ontologies: It is almost always worth considering what someone else has done and checking refinement and extends existing sources for design domain and task. There is no valid reason to expend resources to

build an ontology, which is already available. In some cases, a similar kind of ontology can be derived from the available one. Reusing existing ontologies may be a requirement if the system needs to interact with other applications that have already committed to particular ontologies or controlled vocabularies (Natalya & Mc Guinness, 2001). Many ontologies are already available in electronic form and can be imported into an ontology-development environment that someone is using. The formalism in which an ontology is expressed often does not matter, since many knowledge-representation systems can import and export ontologies. Even if a knowledge-representation system cannot work directly with a particular formalism, the task of translating an ontology from one formalism to another is usually not a difficult one. For instances, the following two ontologies are build using two different tools but one can get the big idea even he/she is not using the same tool.

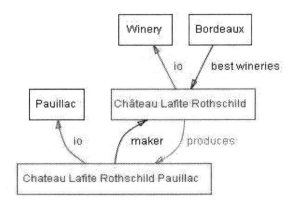

Figure 4.5. Some Classes, Instances, and Relations among Them in the Wine Domain
(Natalya & Mc Guinness, 2001)

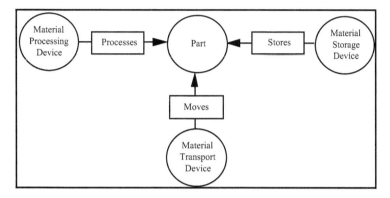

Figure 4.6. Parts–Equipment Relationships

If someone is using the same tool, he/she can import the entire information for his/her own development. There are libraries of reusable ontologies on the Web and in the literature. Stanford University has few online libraries of different domain and site-specific ontologies such as the Ontolingua ontology library (http://www.ksl.stanford.edu/software/ontolingua/) or the DAML ontology library (http://www.daml.org/ontologies/). There are also a number of publicly available commercial ontologies (e.g., UNSPSC (www.unspsc.org), RosettaNet (www.rosettanet.org), and DMOZ (www.dmoz.org)).

For example, a knowledge base of enterprise integration ontology may already exist. If one can import this knowledge base and the ontology on which it is based, he/she will have not only the classification of enterprises but also the first pass at the classification of integration characteristics used to distinguish and describe the terms. Lists of enterprise integration terms may already be available from commercial Web sites that enterprise people consider use to integrate their enterprise.

Step 3. Organize the project: This activity will set different task to be performed to build the new ontology after checking that there are no ontology available to reuse as is or with minor change. Some of the tasks are to form Development Team, break down the tasks, assign team members to specific tasks, etc.

An important initial step in developing an ontology description is the formation of a development team. Each member of the team will perform a particular role in the development effort. Individuals who are involved in the modeling may each fulfill several roles, but each role is dealt with distinctly and should be clearly separated in the minds of the participants. The following are the sample roles assumed by the ontology development project personnel:

i.   Project Leader: This administrative role is responsible for overseeing and guiding the entire ontology development effort.

ii.   Analyst/Knowledge Engineer: Personnel with ontology development expertise who will be the primary developers of the ontology description fill this technical role.

iii.   Domain Expert: This role characterizes the primary sources of knowledge from the application domain of interest. Persons filling this role will provide insights about the characteristics of the application domain that are needed for extracting the underlying ontological knowledge.

iv.   Team Members: All persons involved with the ontology description project.

A Work Breakdown Structure (WBS) is a results-oriented family tree that captures all the work of a project in an organized way. It is often portrayed graphically as a hierarchical tree; however, it can also be a tabular list of "element" categories and tasks or the indented task list that appears in the Gantt chart schedule. Figure 4.7 shows the WBS of building design ontology.

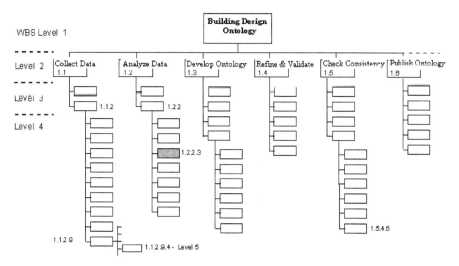

Figure 4.7. WBS of Building Design Ontology

The WBS should be designed with consideration for its eventual uses. WBS design should try to achieve certain goals:

- Be compatible with how the work will be done and how schedules will be managed

- Give visibility to important or risky work efforts

- Allow mapping of requirements, plans, testing, and deliverables

- Foster clear ownership by project leaders and team members and

- Provide data for performance measurement and historical databases

Once a complete WBS is constructed, team members are assigned against each individual task to ensure the progress of the ontology building effort.

Step 4. Collect and Analyze Data: This activity will acquire the raw data needed for ontology development and analyze the data to facilitate ontology extraction. The definition of viewpoint, context, and purpose sets the stage for the data-gathering phase of the ontology captures effort. One of the problems in data collection is determining the appropriate sources of data. Various research experiences indicate that the main data sources are the domain expert and documents relevant to the circumscribed ontology. Regardless of the data collection methods used, it is important at this stage to establish an action plan for collecting data pertinent to the purpose and viewpoint of the model. Once collected, each piece of collected data must be traceable to its source. Traceability of source material is important because it is the data, which provides objective evidence for the basic ontological structures that are later isolated from this data.

Three important support documents can be used to facilitate source data traceability:

i.   Source Material Log: A document that serves as the primary index to all source material used in the project. This log lists all the materials used in the project. This list contains the name of the source materials, sources, who collected and when collected and shown in the Table 4.2. Each entry of this document is then described using separate forms called Source Material Description Form shown in Table 4.3.

Table 4.2. Source Material Log for Building Design Ontology

Source Material Log					
Project: Designing Product/process ontology			**Analysts:** MD Sarder		
Source materia l #	Source material name	Collec ted from	Collecte d by	Date of Collecti on
SM # 1	"Operate a Small Integrated Manufacturing Enterprise" by Don Liles, ARRI, 1998	--------	Sarder	12/15/04
SM # 2	"Product Development and Design for Mfg." by John W. priest & Jose Sanchez, Marcel Dekker, Inc. New York, 2001	--------	Sarder	04/21/04
SM # 3	"Collaborative Evaluation of Early Design Decisions and Product Manufacturability" by S. D. Kleban et el, Proceedings of the 34th Hawaii International Conference on System Sciences - 2001	--------	Sarder	6/23/04
SM # 4	"Complexity and learning behaviors in product innovation" by Ross Chapman and, Paul Hyland. Technovation 24, 2004	--------	Sarder	8/24/04
SM # 5	"Coordination at different stages of product design process" by Antonio J Bailetti et el, R&D Management 28, 4, 1998	--------	Sarder	12/14/04
SM # n	--	--------	---------	---------

Table 4.3. Source Material Description Form for Building Design Ontology

Source Material Description Form
Project: Product & Process Design Ontology
Analyst: MD Sarder
Source material #: SM # 3
Source material name: Collaborative Evaluation of Early Design Decisions and Product Manufacturability" by S. D. Kleban et el, Proceedings of the 34th Hawaii International Conference on System Sciences – 2001
Purpose: To record the relevant source statements that help individuate ontology elements in the product & process design domain.
Comments: This source material concerns early design stage and manufacturing of goods.
Abstract: In manufacturing, the conceptual design and detailed design stages are typically regarded as sequential and distinct. Decisions made in conceptual design are often made with little information as to how they would affect detailed design or manufacturing process specification. Many possibilities and unknowns exist in conceptual design where ideas about product shape and functionality are changing rapidly. Few if any tools exist to aid in this difficult, amorphous stage in contrast to the many CAD and analysis tools for detailed design where much more is known about the final product. The Materials Process Design Environment (MPDE) is a collaborative problem solving environment (CPSE) that was developed so geographically dispersed designers in both the conceptual and detailed stage can work together and understand the impacts of their design decisions on functionality, cost and manufacturability.
Terms Supported: T#2, T#5, T#12, T#15
Statements supported: SS#3, SS#5

ii. Statement Pool: This document records meaningful statements made by different individuals, as well as statements extracted from source documents during the ontology development effort. An example statement about the engineering design is: "Engineering Design activities result in recommended manufacturing specifications that satisfy the customer's functional performance requirements and manufacturing constraints". Table 4.4 shows an example of a source statement pool. Table 4.5 shows source statement description form.

Table 4.4. Source Statement Pool for Building Design Ontology

Source Statement Pool		
Project: Product & Process Design Ontology		**Analysts:** MD Sarder
Source Statement #	**Source Statement**	Supported by
SS # 1	Engineering Design activities result in recommended manufacturing specifications that satisfy the customer's functional performance requirements and manufacturing constraints.	MD Sarder
SS # 2	Resources may be classified as personnel, computer systems, and facilities.	MD Sarder
SS # 3	-- ------------	---------------
SS # n	--	---------------

Table 4.5. Source Statement Description Form for Building Design Ontology

Source Statement Description Form			
Project: Product & Process Design Ontology		**Analysts:** MD Sarder	
Source Statement #: SS#1	Statement #S Evolved To:	Status: *Active* / *Retired*	
Source Material #: SM#2	Statement #S Derived From:	*Original* / *Derived*	
Source Statement: Engineering Design activities result in recommended manufacturing specifications that satisfy the customer's functional performance requirements and manufacturing constraints.		Supported by:	
Version 1: Engineering design deals with the target specifications that will meet customer's functional requirements.		Supported by:	
Version 2:		Supported by:	
Version 3:		Supported by:	
Comments:			

iii. Term Pool: The Term Pool alphabetically records all the meaningful terms relevant to the ontology building effort. Terms would typically connote kinds/instances of kinds, and relations/instances of relations. In this example, a term pool would include such terms as engineering design, embodiment design, rapid prototyping, etc. Table 4.6 shows a sample term pool and Table 4.7 shows the description of each term.

Table 4.6. Term Pool for Building Design Ontology

Term Pool				
Project: Product & Process Design Ontology			**Analysts:** MD Sarder	
Term #	**Term**	Source Statement Reference	Source Material Reference	Support list
Term # 1	Engineering Design	SS#1	SM#2	MS
Term # 2	Embodiment Design	SS#2	SM#13	DS, MS
Term# 3	---------------	-------------------	------------------------	----------------
Term# n	---------------	-------------------	------------------------	-----------------

Table 4.7. Term Description for Building Design Ontology

Term Description Form		
Project: Product & Process Design Ontology		**Analysts**: MD Sarder
Term #	**Term**	**Description**
Term#1	E. Design	Engineering design is process of translating customer requirement into design specifications.
Term#2	Resource	Resources are objects/personnel that are consumed, used, or required to perform activities and tasks. Resources play an enabling role in processes.
Term # 3	-------------	--
Term # n	-------------	--

The objective of data analysis of the ontology development process is to analyze source material that has been collected and construct an initial characterization of the ontology. This task is performed by the knowledge engineer/analyst closely teaming with the domain expert. This task will typically involve the activities such as identify relevant design processes, list the objects of interest in the domain, examine boundary objects for boundary refinement, etc.

i.   Identify relevant design activities

In a product and process environment, the natures of many kinds of things in a domain, especially the important relations they bear to other entities in the enterprise, are revealed not so much by examination of those entities but the roles they play situated in the processes in which they figure. Hence, the first objective in ontology building in product and process design environment is to capture the relevant design activities in which the ontological elements of the domain participate. For example, in a design situation, these will include conceptual design, embodiment design, detail design, design for manufacturing, design for safety, life testing, prototyping, etc. These then provide the necessary contextual information for the construction of accurate and complete domain ontologies.

ii.   List the objects of interest in the domain

Several objects will be fairly obvious from an initial study of the activity descriptions resulting from the previous step. Other objects will be identifiable from the source data such as the Statement Pool and the Term

Pool. For example, the different kinds of design aspects, decision analysis, methods, tools, and fixtures that are associated with product and process design will be obvious ontology candidates for design ontology. The viewpoint and context statements constructed earlier in the development process will guide the level of detail that needs to be employed to develop this list.

iii.   Examine boundary objects for boundary refinement

The initial boundaries defined in the context statement may need to be redrawn to facilitate better conceptual structuring of the ontology. Boundaries are often expanded to accommodate important objects that were earlier on the boundary. For example, consider a Plotter machine that is used in the design processes for ABC's design. Suppose that "drawing equipment" were initially excluded from the scope of the project. Suppose further that there are eleven other kinds of drawing machines that are used to draw and print design drawings at ABC. At this point, the boundaries are redrawn to explicitly include drawing equipment as part of the design ontology.

iv.   Partition the domain into subsystems

Systems are defined as collections of physical and/or conceptual objects that work together for a common purpose. Organizing ontologies by the systems provides a clear conceptual framework for subsequent analysis of ontological knowledge. It is therefore important to partition the focus domain into clearly delineable subsystems early in the ontology development process. This design ontology building method provides a graphical language that supports the conceptual activities such as representation of a system at varying levels of abstraction.

Step 5. Develop Initial Ontology: This activity is to develop a preliminary ontology from the acquired data. In the previous step relevant data was collected and analyzed them to use in this step. In this step there are a series of task according to tool to be selected. For IDEF 5 tool, the task could be identifying Proto Properties, proto Relations, and proto Kinds, classify Kinds, Properties, and Relations, etc.

i.   *Identify Proto Properties and Proto Kinds*

Properties are the characteristics that hold of objects in the real world. Examples of properties are weight, color, age, shape, etc. In this ontology building method, concepts are initially catalogued as "proto" concepts, that is, they are tentative and subject to further inquiry before final change of status by eliminating the "proto" prefix. Thus, potential candidates for "properties" in the ontology are initially called "proto properties." Similarly, there are "proto kinds" and "proto relations" (described later in this section). Proto property identification usually occurs concurrently with proto kind identification. This is because kinds are usually individuated on the basis of the properties that they exhibit. Listing properties is a relatively straightforward task because properties are readily observable and are often measurable.

A *proto kind* is the result of a preliminary attempt at individuating a kind. This task essentially consists of associating the objects identified in the data analysis activity with the proto properties identified. It may be instructive to perform this association process in two stages. First, the association is carried up to the point where the proto kind can be clearly distinguished from any other proto kind, that is, the proto kinds have a basis for being uniquely individuated. Properties that contribute to the uniqueness of a kind are candidate-defining properties. Defining properties stipulate necessary conditions for membership to a kind. Once the defining properties are identified, the remaining properties (non-defining) that are used to characterize the kinds in greater detail are associated with the kinds. At this stage of the analysis, it often becomes clear which proto kinds represent genuine kinds, where two terms have been used to indicate the same kind (namespace redundancy), where the same term is used to indicate distinct kinds (namespace ambiguity), and so on. Once the characterization of a proto kind is relatively complete, it is converted to a kind. That is, classification as a "proto" concept is no longer necessary in view of the evidence that supports the concept.

ii.   *Classify Kinds and Kind Hierarchy*

Deciding whether a particular concept is a Kind in an ontology or an individual instance depends on what the potential applications of the ontology are. In case of design ontology, recognizing the multiplicity of classification mechanisms in different domain areas, this method provides a range of different classification relations to aid domain experts to identify kinds & sub kinds. Depending on the context of use, the subkind (classification) relation can be categorized under three headings as described in the following.

a. Generalization-specialization: The generalization classification *Is-a* relation links a general kind with a specialization of the kind. For example, a *conceptual design* kind is a specialization of a *design steps* kind. This type of classification relation is widely used in a variety of different application domains.

b. Natural kind classification: Ontologies of physical objects are classified using the "kind classification" meaning of the Is-a relation. This relation, often dubbed "a kind of object" (AKO), bestows the distinguished status of kind hood to the related objects. Often there are no necessary and sufficient conditions for entry into a particular kind, and objects just "are" (i.e., by definition), or happen to be, of particular kinds. The AKO relation is used predominantly for classifying natural objects and natural phenomena. For example, a *design software* is a kind of *design resources*.

c. Description classification: Description classification relations are used to define one object kind in terms of another. This type of Is-a relation is particularly useful for describing abstract object kinds. For example, the assertion that "A square is a rectangle" is a concise way of asserting that "a square is a rectangle with four equal sides." Here, the description of the rectangle is used to define the concept of the square, that is, the square description subsumes the description of the rectangle.

Figure 4.8 illustrates the classification of design resources. Typically, design resources can be categorized as collaborative design team, equipment, and tools. Collaborative design team consists of design experts from various disciplines. Equipment can be further categorized as design testing machine, rapid prototype machine, and material storage device. A tool can be design software and decision analysis tool.

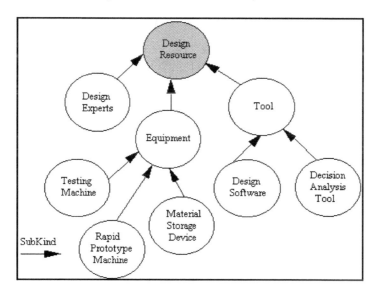

Figure 4.8. Classification of Design Resources

Once kinds are identified, they need to be placed in a hierarchy of order. There are several possible approaches in developing a kind hierarchy (Natalya and Mc Guinness, 2001):

- A top-down development process starts with the definition of the most general concepts in the domain and subsequent specialization of the concepts. For example, one can start with creating kinds for the general concepts of product and process design. Then he/she specializes the design Kind by creating some of its sub kinds: conceptual design, embodiment design, and detail design. One can further categorize the detail design, for example, into part analysis, field-testing, FEMA analysis, and so on.

- A bottom-up development process starts with the definition of the most specific kinds, the leaves of the hierarchy, with subsequent grouping of these kinds into more general concepts. For example, one can start by defining kinds for thermal analysis and electric analysis. He/she then creates a common super class for these two kinds "Stress analysis" which in turn is a sub kind of part analysis.

- A combination development process is a combination of the top-down and bottom-up approaches: One can define the more salient concepts first and then generalize and specialize them appropriately. He/she might start with a few top-level concepts and a few specific concepts and then relate them to a middle-level concept.

iii.   *Identify Proto Relations*

A proto relation is the result of a preliminary attempt at individuating a relation. Proto relations express hypothesized associations between proto kinds or between kinds. The identification and characterization of relations is often the most difficult part of knowledge capture. The identification of proto relations refers to the activity of recognizing the existence of, or becoming attuned to, a particular proto relation in the domain. Characterization follows identification, and refers to the activity of identifying and specifying the properties of a proto relation in a manner that will allow the relational knowledge to be used for making useful inferences at some future time. Thus, recognizing that a tool post is "Above" the lathe bed is the act of discovering and asserting its existence and giving it a name. Characterizing it will involve making assertions such as: the above relation is transitive. Suppose consider the relation between a part to be designed and the different kinds of equipment in the design process. Design drawing tool (such as Auto CAD software) typically require information about the detailed geometry of a part in order to draw the part. However, a decision analysis tool does not require more information than the alternatives. At the same time, the material storage devices require "minimum enclosing box" dimensions and the part weight information to perform the storage function. These associations are now treated as proto relations. The Relation

Schematic shown in Figure 4.9 facilitates the conceptual analysis of this

proto relation.

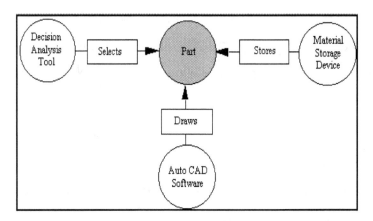

Figure 4.9. Part-Resource Relationships

Figure 4.7 asserts the association of the part with other design resources.

When there is sufficient evidence to justify that these associations are bona

fide relations, the proto-relations are made into relations. This completes the

activity of relation identification. The next step in this ontology capture

method is to "characterize" the relation that specifies "rules of allowable

behavior" for the relation. It tells the importance of relationship. Two types

of relationships are allowed in IDEF 5 tool, they are primary relationships

and secondary relationships. To represent primary relationships, an

arrowhead is placed on the head and to represent secondary relationships,

arrowhead is placed in the tail.

Step 6. Refine and Validate Ontology: This activity will refine and validate the ontology to complete the development process. The refinement process is essentially a deductive validation procedure: the ontological structures are tested with actual data, and the result of the instantiation is compared with the ontology structure. If the comparison produces mismatches, every such mismatch must be adequately resolved. Ontology refinement includes the following steps.

i.   *Kind Refinement Procedure*

The kind refinement procedure is summarized in the following steps:

a. Make instances of the kinds (and proto kinds). The examples may be constructed from the available source data (source data catalog), otherwise new data must be gathered for the purpose of constructing these examples. The examples must be reasonably representative, with at least one exception case included, if possible. Each of the (proto) kind instances created is populated with properties. Classification diagrams and kind specification forms are used to support the kind instantiation process.

b. Record information that cannot be recorded in the kind instances. Determine whether this additional information is really necessary, and if so refine the structure of the kind to include the information.

c. Check whether two instances of the same kind have different defining properties. In such cases, check whether the viewpoints are different. If not, the inconsistencies will have to be resolved by refining the ontology (for instance, by redefining the contentious property to be non-defining).

ii.   *Relation Refinement Procedure*

The relation refinement procedure is summarized in the following steps:

a.   Make instances of the relations (and proto relations). The examples may be constructed from the available source data (source data catalog), otherwise new data must be gathered for this purpose. The ontology relation diagrams and the relation specification forms are used to aid the instantiation and validation procedure.

b.   The properties of each of the relation instances are compared with the properties identified in the ontology description, and any mismatches are resolved. Moreover, check for missing relation properties, and add them if needed.

c.   Sample instances of selected relations. Check whether two or more instances of such relations are incompatible. For example, one relation says that a fastener must have a sealant and another may say that it cannot have a sealant. Such inconsistencies may be either due to hidden viewpoint differences not recorded in the ontology, or because of differing viewpoints. Incompatibilities that occur because of differing viewpoints may be resolved by splitting the focus relation into different relations, one for each viewpoint. Otherwise, a consensus must be reached to resolve the incompatibility through discussions with the domain expert.

d.   Detect new relations discovered by example that are not captured in the ontology. Add such relations to the ontology.

Step 7. Check consistency & accuracy of ontology: This activity will ensure the accuracy and consistency of captured knowledge. Once ontology building is complete, it is necessary to check the consistency and accuracy of information captured specially for design ontology. This consistency check can be done by the help of consistency matrix, which was developed earlier. The detail steps of the development of consistency methodology are discussed in the end of this chapter. If consistency check comes acceptable, the ontology will be ready to use.

Step 8. Collect additional data and analyze data: If consistency check comes unacceptable, further data collection and analysis will be conducted to resolve the disputes. On the basis of this additional data initial ontology will be developed and refinement will be conducted. Finally consistency will be checked again until it meets the acceptable consistency.

Step 9. Incorporate lessons learned and publish ontology: This will add new findings and new research in the ontology and make available for others to use. Ontology must be dynamic and updated in terms of information content. It must be capable to incorporate new findings and lesions learned and publish in the online ontology repository for others. Figure 4.10 shows the structure of building design ontology.

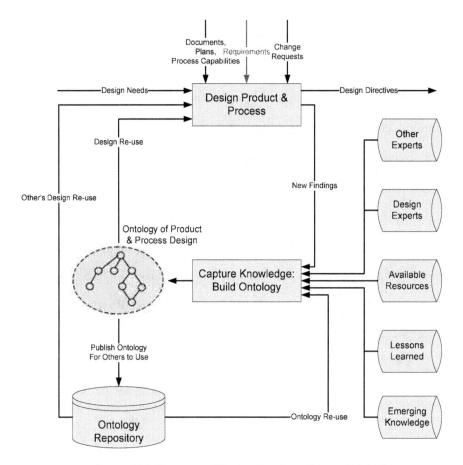

Figure 4.10. Structure of Design Ontology Capture Method

Although these steps of design methodology are listed sequentially, there is a significant amount of overlap and iteration between the activities. Thus, for instance, the initial ontology development (Step # 5) often requires the capture of additional data and further analysis (Step # 4). Each of the nine activities will involve other activities and tasks.

4.8 Methodology for Consistency Matrix

As mentioned earlier that product process design plays a very important & critical role for manufacturing enterprises to survive in the market place. Among all functions of manufacturing enterprise, product and process design is the single most important function. To make sure that the design process is accurate as well as design information is consistent, it is necessary to check the consistency of captured information. The design ontology must posses some/ all of the following characteristics;

- Accurate and up-to-date information

- Easily sharable & reusable

- Enable to capture future research and lesions learned

- Consistent information

- Visual & structured representation

To ensure accurate and consistency of information in the ontology, a consistency matrix was developed using the following methodology. This methodology is straightforward and easy to use for manufacturing enterprises. It has the following steps.

1. Identify the design activities with relative importance in the domain

2. Prioritize design activities of the domain

3. Check captured design activities against the domain design activities

4. Calculate consistency index

Detail description of these steps is discussed in the following sections.

Step 1. Identify all possible design activities with the relative importance in the related domain of product and process design. If one is building design ontology for the design of a sports utility vehicle, he/she needs to find the design activities in the domain of automobile design. Identifying domain design activities can be done using the following steps.

    a. Check for available domain ontology for that particular domain or similar domain and use them appropriately.

    b. Search the literatures, best design practices which, includes journal articles, books, conference proceedings, web sites, etc for the design activities.

    c. Interview related domain experts or design specialists associated with products and process design.

    d. Attend conferences, symposiums, industry group discussions, etc related to product and process design.

Step 2. Prioritize the design activities found in step 1. This prioritization is on the basis of relative importance of the design activities in the design process. This provides an important glimpse to the ontology author about what to include in the ontology and what is not. Prioritization can be done using ABC analysis or Pareto analysis. Pareto analysis (sometimes referred to as the 80/20 rule and as ABC analysis) is a method of classifying items, events, or activities according to their relative importance (Balling, Richards, 2000). It is frequently used in inventory management where it is used to classify stock items into groups based on the total annual expenditure for, or total stockholding cost of, each item. But it can be used in situation where prioritization is the main task. Organizations can concentrate more detailed attention on the high value/important activities. A Pareto analysis for prioritizing design activities consists of the following steps.

a. List all design activities of the domain of interest

b. Enter relative importance of each activity

c. Calculate the percentage of total importance represented by each activity.

d. Rearrange the list. Rank items in descending order by total value, starting at the top with the highest value.

e. Calculate the cumulative percentage of the total value for each item at the top; add the percentage to that of the item below in the list.

f. Choose cut off points for A, B and C categories.

g. Present the result graphically. Plot the percentage of total cumulative value on the Y-axis and the item number in X-axis. A typical Pareto curve for a design situation is shown in figure 4.11.

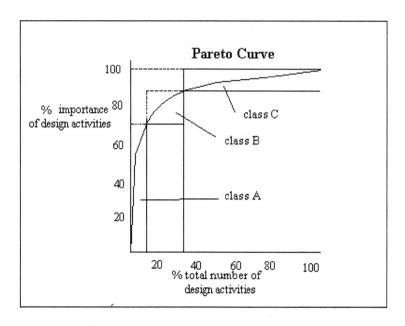

Figure 4.11. Pareto Curve of a Design Situation

Step 3. Check captured design activities against the domain design activities found in step 1 and 2. This can be done with the help of a matrix shown in figure 4.12. The columns of this matrix represent the type A, type B, and type C activities of the domain and the rows of the matrix represent the captured design activities of design ontology, which are under construction.

Captured Design Activities in the Design Ontology		Design Activities																											
		Type A					Type B								Type C														
		Activity #1	Activity #2	Activity #3	Activity #4	Activity #5	Activity #6	Activity #7	Activity #8	Activity #9	Activity #1	Activity #1	Activity #1	Activity #1	Activity #1	Activity #1	Activity #1	Activity #1	Activity #1	Activity #1	Activity #2	Activity #2	Activity #2	Activity #2	Activity #2	Activity #2	Activity #2	Activity #2	Activity #2
Industrial	Activity # 1																												
	Activity # 2																												
	Activity # 3																												
	Activity # 4																												
	Activity # 5																												
Conceptual	Activity # 6																												
	Activity # 7																												
	Activity # 8																												
	Activity # 9																												
	Activity # 10																												
Embodiment	Activity # 11																												
	Activity # 12																												
	Activity # 13																												
	Activity # 14																												
	Activity # 15																												
	Activity # 16																												
Detail	Activity # 17																												
	Activity # 18																												
	Activity # 19																												
	Activity # 20																												
	Activity # 21																												
	Activity # 22																												
Process	Activity # 23																												
	Activity # 24																												
	Activity # 25																												
	Activity # 26																												
	Activity # 27																												
	Activity # 28																												
Raw Score																													

Figure 4.12. Consistency Check Matrix

This matrix counts the number of checks in each area of type A, type B, and type C. Bottom line of the matrix gives the sum of theses counts under each category. This raw score is used to calculate the index in the next step.

Step 4. Calculate the consistency index. This index provides a clear picture of the accuracy and consistency of design information, which is captured to construct the ontology. Index calculation can be done using the following

steps.

a.   Normalize the raw score. From the consistency cheek matrix in

step 3, three different raw scores are found; they are raw score for type A

activities (R_A), raw score for type B activities (R_B), and raw score for

type C activities (R_C). Normalization of these scores can be done using

the following equations.

$$N_A = {R_A}\big/{M_A} \, X100 \text{ ; where } N_A \text{ represents the normalized score for type}$$

A activities and M_A represents the maximum raw score for type A

activities.

$$N_B = {R_B}\big/{M_B} \, X100 \text{ ; where } N_B \text{ represents the normalized score for type}$$

B activities and M_B represents the maximum raw score for type B

activities.

$$N_C = {R_C}\big/{M_C} \, X100 \text{ ; where } N_C \text{ represents the normalized score for type}$$

C activities and M_C represents the maximum raw score for type C

activities.

b.   Calculate the index. Pareto analysis is based on Pareto principle,

which says that typically, type A accounts the first 20% of activities in

the list will account for approximately 70% of cumulative importance.

The next type B accounts 30% of activities, will, typically, account for a

80

further 15% of cumulative importance. These can be subject to less precise control methods. The last type C accounts for the rest 50% of (low importance) activities then account for a mere 15% of importance and can be controlled with a simple system. From this basic principle, the relative weight of each type of activities can be presented as W_A =0.7, W_B =0.15, and W_A =0.15, where W_A, W_B, and W_C are the relative weight of type A, type B, and type C activities respectively. Hence the Consistency Index (CI) of design ontology will be as follows.

$$CI - W_A N_A + W_B' N_B + W_C' N_C$$

$$\text{or } CI = W_A \frac{R_A}{M_A} X100 + W_B \frac{R_B}{M_B} X100 + W_C \frac{R_C}{M_C} X100$$

The result of this index will be a numeric value between 0 and 100. The higher the value, the more consistent the ontology is. If the value of this index is low for a design ontology, there is question of validity of such ontology model.

PART III
A GENERIC PRODUCT & PROCESS
DESIGN ONTOLOGY

CHAPTER 5

PRODUCT AND PROCESS DESIGN ONTOLOGY

5.1 Overview

Design ontology construction is the task of acquiring the terminology, statement structure, and sanctioned inferences of a given products and processes design domain and storing it in a usable representational medium. Ontology includes much more that just "dictionary" information. It seeks to identify the primary entities, or kinds, of objects that are within the domain by isolating the properties that define the members of those kinds and the characteristic relations that hold between domain objects. Building design ontology shows how to capture related domain knowledge, processes captured knowledge to identify entities and relationships, and present the captured information in a way, which is easy to comprehend and easy to share. From the above discussion, this design ontology has three fundamental components. They are as follows.

i. Terminologies: These are the terms and statements associated with products and processes design. This includes a comprehensive list of design terms with definitions if not all terms of products and processes design. Definition of terms not only includes usual meaning but also the contradictory definition, which is later, used to refine the ontology. All terminologies related to products and processes design, identified in this effort are identified and presented in glossary of terms.

ii.   Entities: An entity is an objective category of objects that are bound together by a set of properties shared by all and only the members of the kind. Entities, classes, types, and kinds all indicate some grouping of individuals into categories and roughly equivalent. In the context of design ontology, Entities are the important design types, processes, resources, equipment, tools, etc. All entities related to products and processes design, identified in this effort are identified using IDEF5 and presented in the Appendix A1. Detail descriptions of entities are discussed later in this chapter.

iii. Relations: Relations are the interactions of entities and instances to each other. IDEF5 supports the description of relationships among entities, among instances, and among entities and instances. It explicitly supports the relation "subkind-of". For example, the relation "consists of", is a general feature that holds between a collaborative design team and the multi-disciplinary experts. The relations in an ontology are typically binary; that is to say, they hold between two entities, as with the relation consists of. However, there is no theoretical bound on the number of arguments of a relation; the relation between, for instance, holds between three objects. In this effort, more than forty relations are identified and presented in the Appendix A2. Detail description of relations is discussed later in this chapter.

5.2 Design Ontology

Ontology building is a time consuming and tedious effort. In the following sections, a design ontology of products and processes is constructed using the methodologies already developed in chapter four. This construction process includes determining scope of the design domain, data collection and analysis, identifying ontology components, refining those components, and presenting for easily understanding and uses. All the above-mentioned steps are discussed in the following sections.

5.2.1   Determine the Domain and Scope

Product and process design is a very complex, innovative, and iterative process. Though, it follows structured steps of processes, it varies from product to product. In case of high-end electronic products, conceptual design and testing of products are more important than that of a matured product such as automobile. Product and process design includes industrial design, engineering design, and production design (Rudolph, 2005). It is a part of product development life cycle and product development is again a part of product life cycle. A product life cycle consists of the following steps. Figure 5.1 shows these steps with further decomposition of product development.

      i.   Product development

     ii.   Production

   iii.   Distribution

   iv.   Service

v.   Disposal

Product development is decomposed into sales & marketing and product and process design. As mentioned earlier that product and process design consists of industrial design, production design, and engineering design. In this book, products and processes design was considered as the domain of interest.

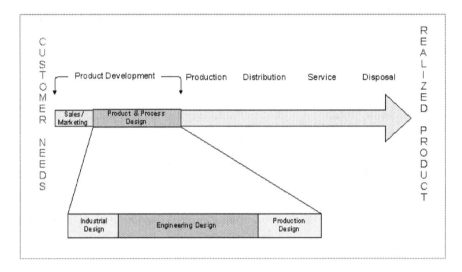

Figure 5.1. Product Life Cycle Phases (Modified from Rudolph, 2005)

Product and process design itself is a huge domain. It is almost always nice to capture all available entities and relations in the domain but sometime impossible and long-term project. The scope of this domain was limited to a generic product and process design with respect to collaborative design team.

5.2.2 Data Collection and Analysis

One of the major and time-consuming tasks of this effort was to collect data related to the domain of interest. The method chosen to collect data was determined from an extensive literature search to determine what tools are available that will get at the question asked. The methods to collects the data was reliable and consistency was maintained.

The data collection method influences a number of factors. The most important ones include survey design: the selection of experts, selection of questions, and the cost of data collection. The data collection method also has a bearing on the timetable of the survey and on the quality of the final results. All these features are considered to design the survey questionnaire and in expert selection process.

Three methods of data collection techniques are used in this discussion. They are literature reviews, interviewing design experts, and attending product and process design sessions in national and international conferences. An extensive literature reviews were conducted to identify product and process design related terminologies, statements, relative importance of design steps, new principles in product and process design, and so on. More than 100 journal articles, conference proceedings, books, white papers, etc are documented as a source of data collection. Survey questionnaires are designed to collect design information from the design experts. A sample questionnaire is presented in the Appendix B1.

More than two hundred terminologies are identified from this data collection effort. A partial snapshot of this data is shown in the following figure.

Anticipatory failure
determination
Assembly drawings
Axiomatic design
Backtracking
Benchmark
Bending
Beta prototype
Beta testing
Bill of materials
Blanking
Blow molding
Boring
Brazing
Built-in-self-test
Bulk deformation
Casting
Clearance fit
Coefficient of thermal
expansion
Collaborative design
team
Company requirements
Component
decomposition
Compression molding
Computer-aided design
Computer-aided
engineering
Configuration design
Configuration
requirements sketch
Consensus decision
Contingency design
Correlation ratings
matrix
Customer requirements
Decision-making
process
Design analysis
Design compatibility

Design for testability/
inspect ability
Design for
serviceability
Design for cost
Design for lifecycle
cost
Design for performance
Design for reliability
Design for safety
Design for assembly
Design for disassembly
Design for adjustability
Design for assembly
Design for close fit
Design for manufacture
Design for the
environment
Design for the extreme
Design for x
Design goal
Design guidelines
Design method
Design need
Design phase
Design problems
Design process
Design requirements
Design reviews
Design validation
Design variable
Design verification
Design-project report
Detail design
Detail drawings
Diagram
Die casting
Drawing
Drilling
Early supplier
involvement

Embossing
Engineering analysis
Engineering change
Engineering design
Engineering design
specification
Equipment
Evaluation
Evaluation criteria
Extrusion
Facing
Fail-safe design
principle
Failure analysis
Feasible design
Field-testing
Finishing
Fixture
Forging
Forming
Formulation
Function decomposition
Functional test
Grinding
Hazard analysis
Highly accelerated life
test
Honing
Human factors
Industrial design
Injection molding
Insertion
Investment casting
Lapping
Machine
Machine tool
Machining
Manufacturing Design
Mechanical press
Milling
Mistake-Proofing

Figure 5.2. Partial Term Pool of Design Domain

5.2.3 Initial Design Ontology

Once term pool is compiled, data analysis begins. In this book data analysis is conducted in accordance with IDEF5 methodology. A list of proto-kinds and a list of proto-relations are constructed using the top down approach, mentioned in earlier chapter. Using IDEF5 analogies and consistency check, a final list of kinds/ entities and relations are derived. In the following section, a detail description of all identified entities/kinds and relations are presented.

5.2.3.1 Entities of Design Ontology

The elements described in this section are the entities of the domain of products and processes design. Identification of these design entities was done using a well-structured method of IDEF5 modeling tool. Two major types of entities are identified in this book. They are design activity and design object. Since the domain of interest is products and process design, which itself an activity, majority of entities in this effort fall under this category. Design activities are any process of transforming state of the environment that happens over time. Other type is considered as resources, mechanisms, controls, inputs, and outputs of the design processes. An object is defined as a tangible or conceptual entity in the design domain. Object is further classified into two sub categories; they are Actor and Products. Actor could be a design expert, collaborative design team, or other entity capable of actively participating in an activity. Actor is classified into three sub categories. The taxonomy of the design entity is shown in figure 5.3.

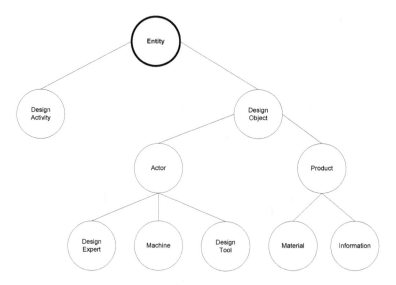

Figure 5.3. Taxonomy of Design Entity

A design expert is defined as a human being entity that performs activities to achieve design goals. Its ability to actively participate in the planning process is what separates a Design Expert from a Machine. A machine is a non-human entity, which has the capacity to carry out design functions. Design Tool is also known as a non-human entity, which has the capacity to carry out design functions. Product category of entities in the design domain is considered as any passive object that can be used, created, or modified by an activity in the design domain. These are typically the inputs and outputs of a design process. Product entity has Material and Information type of design entities. Material is an object that can be used, created, or modified or consumed by an activity in the design domain. (e.g. Raw materials, Subassemblies). Information is defined as an object that can be used, created, or modified by an activity and whose essence is to act as information in the design domain.

Each entity of the products and processes design domain has list of attributes and defining properties, which makes them distinguishable. Table 5.1 shows detail description of a design activity type entity called "Product Design". IDEF5 supports this form of representation, which is hierarchically structured.

1.1 PRODUCT DESIGN	
A process of transforming customer/design requirements into detail design specification using design resources while satisfying design constraints.	
Attribute	**Description**
Name	Product design
Number	A unique qualifier for the activity
Parent	Design Activity
Documentation	User defined description of the Product design
User-Attributes	User defined attributes for Product design
Components	ACTIVITIES that make up this Activity. Such as design analysis, test and evaluation, validation and verification, etc.
Authorizer	The agent responsible for this ACTIVITY, usually the collaborative design team is considered the process owner and has authority over design process.
Objects	The objects (non resource) used by this Activity, Objects in this case are material and information.
Resources	The collaborative design team, machine, or other equipment that performs the Activity
Sub kinds	Formulation, Conceptual design, Embodiment design, and Detail design.

Table 5.1. Product Design Entity

2.1.1 MATERIAL
Any material object that can be used, created, or modified or consumed by an activity in the design domain (e.g. Raw materials, Subassemblies).

Attribute	Description
Name	Material 's name
Number	A unique qualifier for the Material
Parent	Product
Documentation	User defined description of the Material
User-Attributes	User defined attributes for a Material
Performs	The ACTIVITIES, which this MATERIAL is used to perform.
Sub kinds	Component of kind

Table 5.2. Material Entity

Another example of an object category entity called "Material" is shown in Table 5.2. This type of representation scheme gives a clear description of entity. In appendix A1, all entities of the design domain, identified in this effort are presented in the order of hierarchical relationship. In the following sections, some of these identified entities are further discussed and presented in a visual form, which is supported by IDEF5 representation scheme. In some cases a modification was made to draw figures to accommodate additional information. Entities in the domain do not mean anything without the relationships among each other. In the immediate following section, a detail description of such relations in the design domain is presented.

5.2.3.2 Relations of Design Ontology

Relationship is the way to connect, associate, or relate two or more entities with each other (Uschold *et al.*, 1995). It defines the interaction of entities. In products and processes design, this relation describes how the entities are associated and how the overall design function is performed. Figure 5.4 shows different relations with Parametric Design entity and with other design entities.

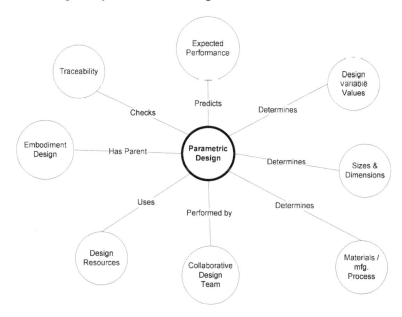

Figure 5.4. Relations of Parametric Design Entity with Other Entities

Each of these relations shown in the figure 5.4 is described individually using the IDEF5 format. Table 5.3 shows an example relationship specification for relation "Performed by" of figure 5.4.

Table 5.3. Example Relationship Specifications

1. PERFORMED BY		
Entity 1	**Entity 2**	**Definition**
Parametric Design	Collaborative Design Team	A person, group of person, or machine performs the activity "Parametric Design"

IDEF5 allows the differentiation of relationships according to the importance in the design processes. Two types of relationships are first order or primary relationships and second order or secondary relationships. Primary relationships are represented with an arrow in the head and secondary relations are presented with an arrow in the tail. In figure 5.5, component of systems design for a matured chemical product is described, where manufacturing process design, quality design, testability design, safety design, and disposal design have first order relationships with the systems design and the rest have secondary relationships with the systems design.

This type of relationships changes from industry to industry, product to product, and company to company. For example, electronic industry requires more importance on product development and testing whereas a pharmaceuticals company emphasizes on safety and quality of product. All relations related to the products and processes design domain, identified in this effort, are presented in Appendix A2.

Figure 5.5. Components of Systems Design for a Matured Chemical Product

5.2.3.3 Descriptions of Design Entities

This section describes the representation scheme for some entities of design domain. It provides an elaborate description of entities, its components, and relations with other entities. As mentioned earlier, that product and process design is a series of sequential activities, which transform customer's requirements into production release with detail engineering specification and detail process specifications. To comprehend the whole design process, a complete picture of a generic product and process design is presented with the help of process diagrams n the very end of this section.

5.2.3.3.1 Industrial Design & Formulation

These are processes of transforming customer/design requirements into design requirements using design resources and satisfying design constraints. This activity ensures feasibility of the design. A detail description is shown in figure 5.6.

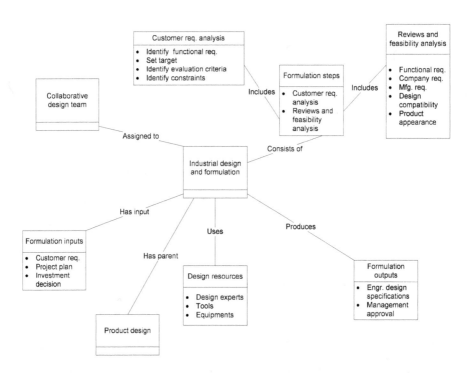

Figure 5.6. Industrial Design and Formulation

5.2.3.3.2 Concept Design

Concept Design is an abstract embodiment of a working principle, geometry, and material; a phase of design when the physical principles are selected. A detail description is shown in figure 5.7.

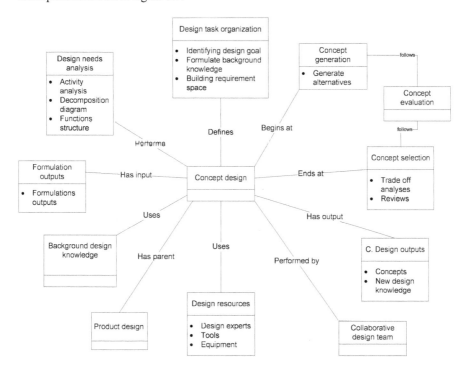

Figure 5.7. Concept Design and Its Relationships with Other Entities

5.2.3.3.3 Embodiment Design

It is a process of selecting detail specifications of product and it's part with the consideration of various aspect of design such as safety, manufacturability, assembly, environment, etc. It has two sub kinds/ entities; configuration design and parametric design. Both configuration design and parametric design has their own sub kinds. A detail description is shown in figure 5.8.

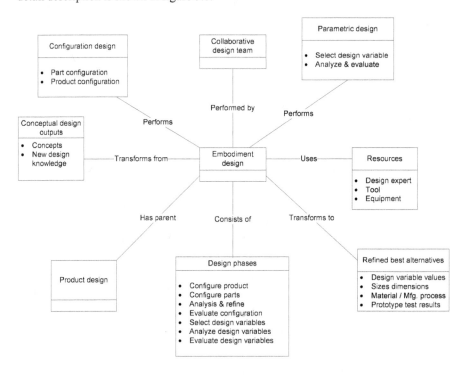

Figure 5.8. Embodiment Design Process

5.2.3.3.4 Detail Design

Detail Design is a phase of design that results in the preparation of a package of information that includes drawings and specifications sufficient to manufacture a product. A detail description is shown in figure 5.9.

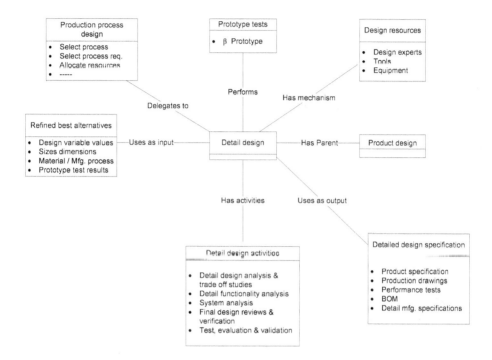

Figure 5.9. Detail Design Schematic

5.2.3.3.5 Production Process Design

Defining and designing a manufacturing process to accommodate the specific requirements of a given product while, meeting process quality and cost objectives. A detail description is shown in figure 5.10.

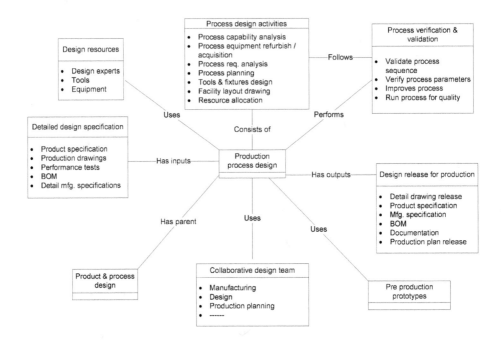

Figure 5.10. Production Process Design Entity

5.2.3.3.6 Analyze Design Needs

It is a process of identifying and clarifying product's functional requirements in detail level. A detail description is shown in figure 5.11.

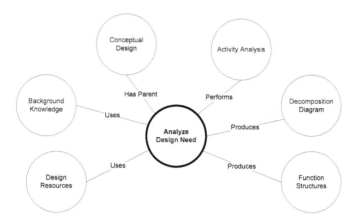

Figure 5.11. Analyze Design Need

5.2.3.3.7 Generate Alternatives

It is a process of creating alternatives or design concept, which is the earliest representation of a new product or of alternative approaches to designing a new product. A detail description is shown in figure 5.12.

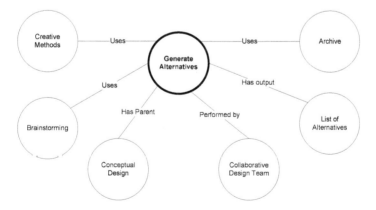

Figure 5.12. Generating Alternatives

5.2.3.3.8 Conceptual Design Analysis & Refine

These are processes of determining concepts/ alternatives considering all aspect of design principles and review those alternatives to refine. A detail description is shown in figure 5.13.

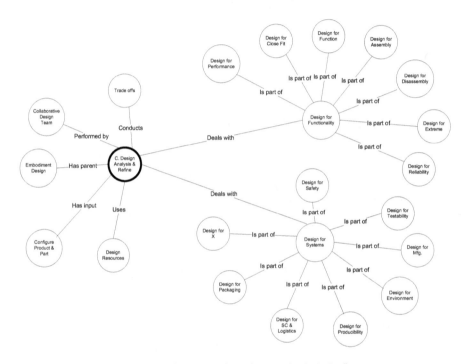

Figure 5.13. Conceptual Design Analysis & Refine

5.2.3.3.9 Concept Evaluation & Selection

These are processes of evaluating all generated alternatives to find out the feasible and producible alternatives and predict their performances. They select the best alternatives among available design alternatives. A detail description is shown in figure 5.14.

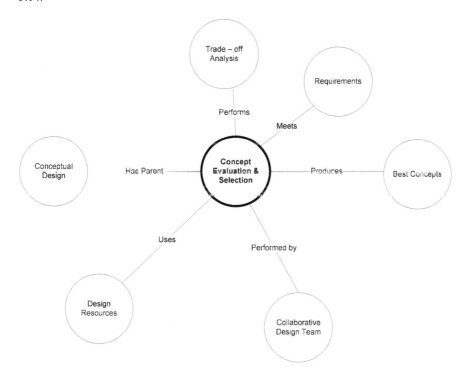

Figure 5.14. Concept Evaluation and Selection Process

5.2.3.3.10 Configuration Design

Configuration Design is a process of selection and arrangement of features on a part; or the selection and arrangement of components on a product; a phase of design when geometric features are arranged and connected on a part, or standard components or types are selected for the architecture. A detail description is shown in figure 5.15.

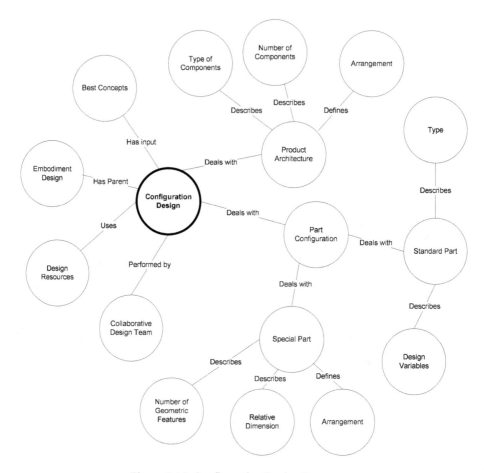

Figure 5.15. Configuration Design Process

5.2.3.3.11 Parametric Design

It is a phase of design that determines specific values for the design variables. A detail description is shown in figure 5.16.

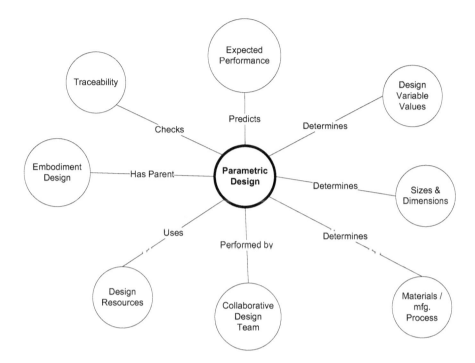

Figure 5.16. Parametric Design Entity

5.2.3.3.12 Detail Design Analysis

It is a phase of design that predicts or simulates performance of each alternative, reiterating to assure that all the candidates feasible. A detail description is shown in figure 5.17.

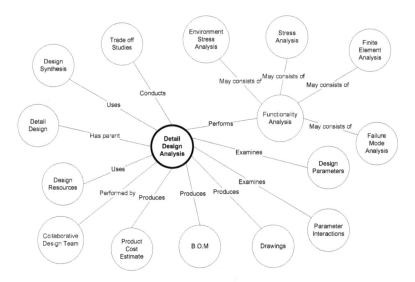

Figure 5.17. Detail Design Analysis Process

5.2.3.3.13 Design Reviews & Verification

These are phases of design that identifies technical performance, risks, improvement of performance & process, and verifies design parameters. A detail description is shown in figure 5.18.

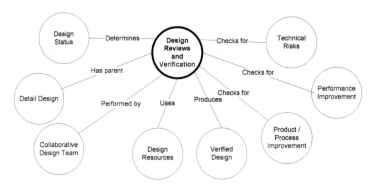

Figure 5.18. Design Reviews & Verification

106

5.2.3.3.14 Test & Validation

These are processes of evaluating each design specifications and validation through extensive testing and simulation. A detail description is shown in figure 5.19.

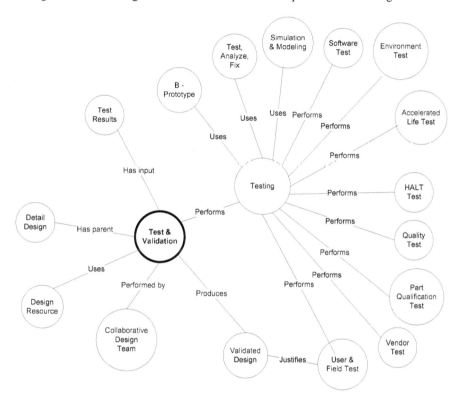

Figure 5.19. Test & Validation in Detail Design Process

5.2.3.3.15 Detail Systems Design

It is a phase of design that determines the required systems specifications for the product to be produced. A detail description is shown in figure 5.20.

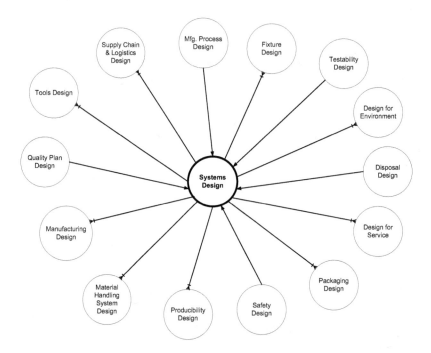

Figure 5.20. Detail Systems Design

5.2.3.3.16 Production Process Design Analysis

Defining and designing a manufacturing process to accommodate the specific requirements of a given product while, meeting process quality and cost objectives. It includes production planning, which is a process of planning and determining process specifications to produce the designed products. A detail description is shown in figure 5.21.

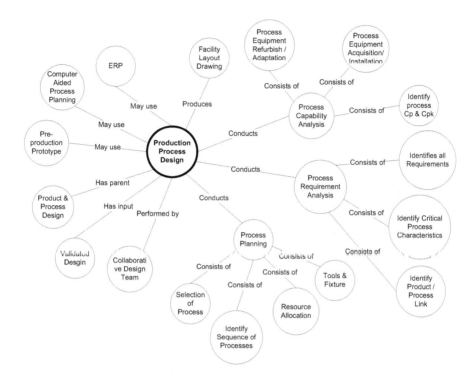

Figure 5.21. Production Process Design Analysis

5.2.3.3.17 Design Resources

These are mechanisms or inputs in the design process. They can be design experts, equipments, or design tools. A detail description is shown in figure 5.22 and component of design resources is shown in figure 5.23.

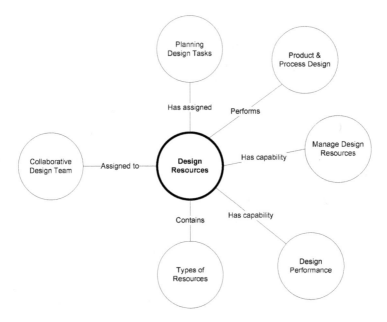

Figure 5.22. Design Resources

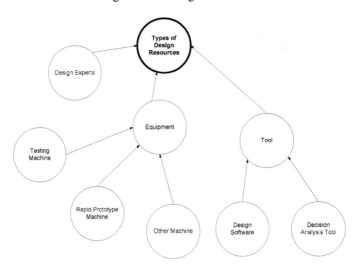

Figure 5.23. Components of Design resources

5.2.3.3.18 Collaborative Design Team

A multidisciplinary design team consisted of people including management, designers, product support, vendors, and customers. The key objective is to improve communication in the design process. A detail description is shown in figure 5.24.

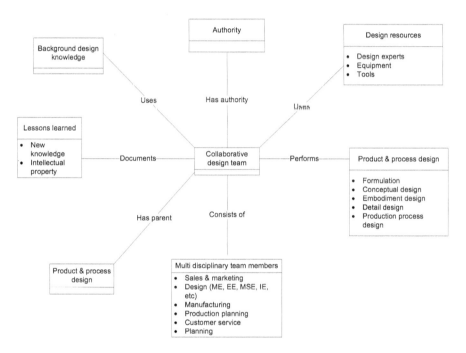

Figure 5.24. Collaborative Design Team Formations

5.2.3.3.19 Design Constraints

These are obstacles to design products and processes. They can be financial, technical or administrative in nature. A detail description is shown in figure 5.25.

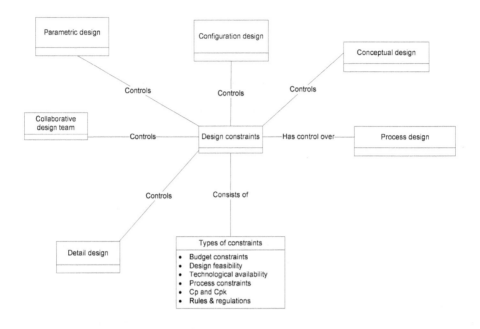

Figure 5.25. Design Constraints

5.2.3.3.20 Prototypes in the Design Processes

Prototypes are physical models or representations of the new product concept or design. Depending upon the purpose, prototypes may be non-working models or representations, functionally working, or both functionally and geometrically complete and accurate. Prototypes (physical, electronic, digital, analytical, etc.) can be used for the purpose of, but not limited to: a) assessing the feasibility of a new or unfamiliar technology, b) assessing or mitigating technical risk, c) validating requirements, d) demonstrating critical features, e) qualifying a product, f) qualifying a process, g) characterizing performance or product features, or h) elucidating physical principles. Different type of prototypes is shown in figure 5.26.

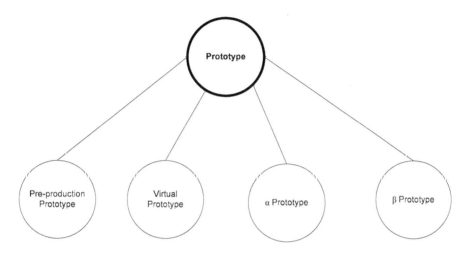

Figure 5.26. Kinds of Prototype in the Design Process

5.2.3.3.21 Technical Requirements

These are product-specific performance and functional characteristics based on analyses of: customer needs, expectations, and constraints; operational concept; projected utilization environments for people, products, and processes; and measures of effectiveness. A detail description is shown in figure 5.27.

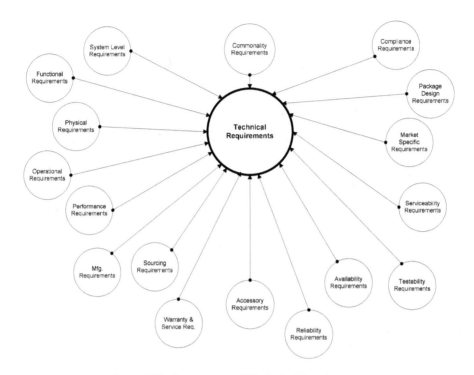

Figure 5.27. Components of Technical Requirements

5.2.3.3.22 Deriving Activity Diagram (IDEF0)

Ontology model (IDEF5) is the complete representation of the product and process design. One can retrieve relevant information from an ontology and draw activity diagram using IDEF0. Figure 5.28 shows an example of deriving IDEF0 Diagram from Ontology Model (IDEF5) of formulate design problem.

114

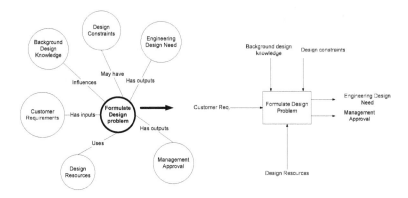

Figure 5.28. Deriving IDEF0 Diagram from Ontology Model

5.2.3.3.23 Deriving Process Diagram (IDEF3)

Similar way, mentioned earlier, one could derive a relevant process diagram (IDEF3) from an ontology model. Figure 5.29 shows a conversion of an ontology model into a process model.

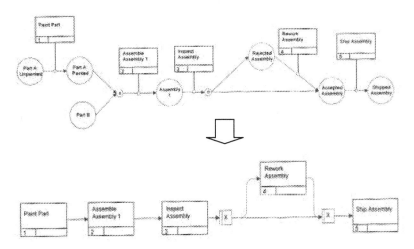

Figure 5.29. Deriving IDEF3 Diagram from Ontology Model (Presley, 1997)

5.2.3.3.24 Hierarchy & Sequence in the Design Process

Entities in the design domain are hierarchically structured. A partial Hierarchy of Product & Process Design is shown in figure 5.30.

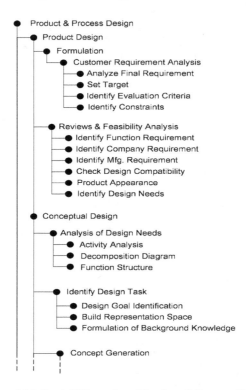

Figure 5.30. Partial Hierarchy of Product & Process Design

Product & process design starts at industrial design & formulation and ends at production process design. There are five major steps of product and process design identified through ontology development process. Each step has several sub steps. All these steps and sub steps are sequential and dependent on the predecessor steps. Detail descriptions of each design steps are shown in figure 5.31 through figure 5.37.

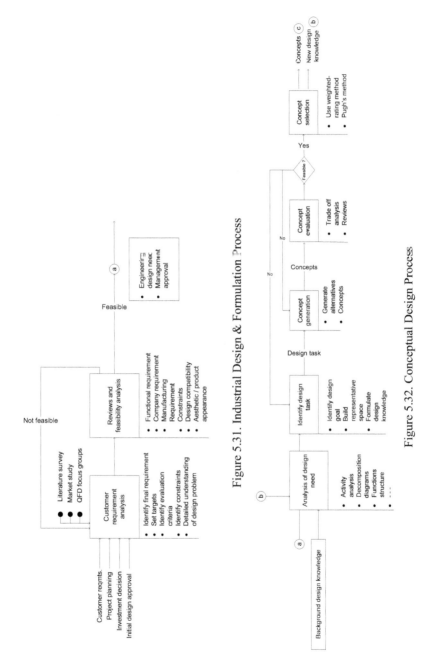

Figure 5.31. Industrial Design & Formulation Process

Figure 5.32. Conceptual Design Process

117

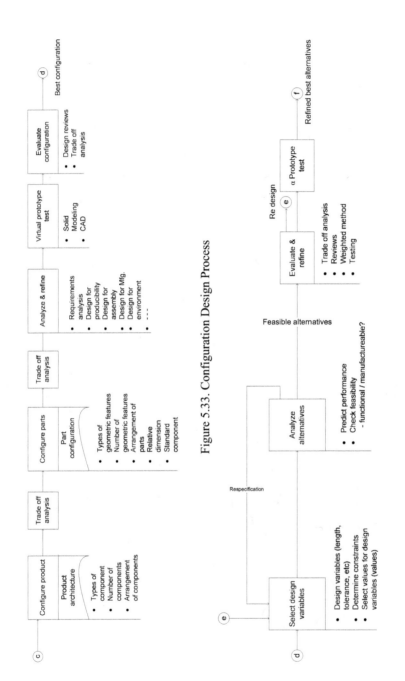

Figure 5.33. Configuration Design Process

Figure 5.34. Parametric Design Process

118

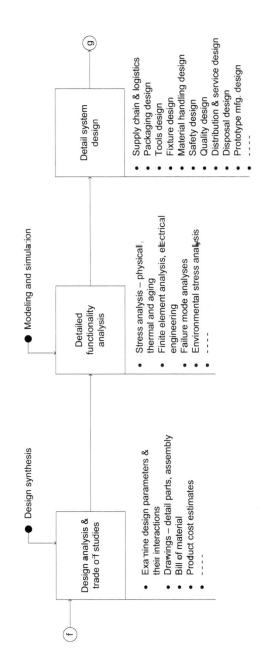

Figure 5.35. Detail Design Process

119

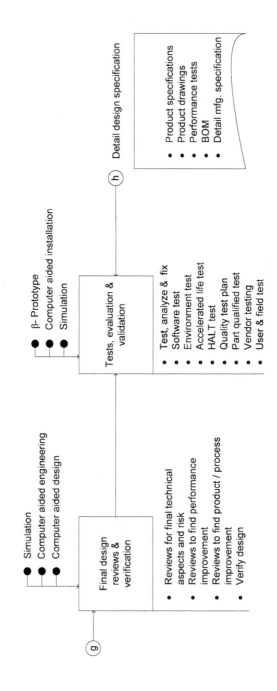

Figure 5.36. Detail Design Process Final Stages

120

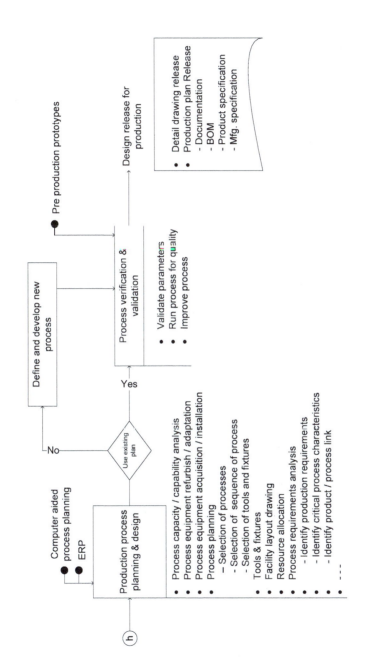

Figure 5.37. Production Process Design

121

5.2.4 Refine & Validate Ontology

The entities, properties, attributes, and relations defined and developed in the previous sections are refined and validated using the IDEF5 methodology. This activity will affirm the validity of the design ontology to the products and processes design domain. Typically this step will identify the need to reiterate some of the steps accomplished earlier. It resolves the contradictory concepts or definition of design terms in the ontology. The outputs of the previous steps are definitely pieces of ontology. As mentioned earlier, ontology has three basic things. They are terminologies, entities, and relations. For example, terminologies are similar to dictionary, which define the terms of the domain. Entities/kind and sub kinds are the building block of an ontology. Each entity has list of attributes and defining properties, which are unique and separate from other entities. Relations are also very important building block within the ontology. It provides ontology semantics, which is absent in the dictionary.

Strictly speaking, validation of ontology was not performed in this effort. Validation is a process of checking whether the ontology is working or mimicking the reality. To validate an ontology, it is necessary to implement in a number of companies and check for consistency. In this book, only two companies were selected for implementation, which is considered as a demonstration not a validation process. Checking consistency of information is another way of validating ontology. In chapter 6, consistency matrix developed in the previous chapter was implemented. As part of the refinement and validation process, a complete ontology model of products and processes design was created and the whole top-level ontology is shown in figure 5.38.

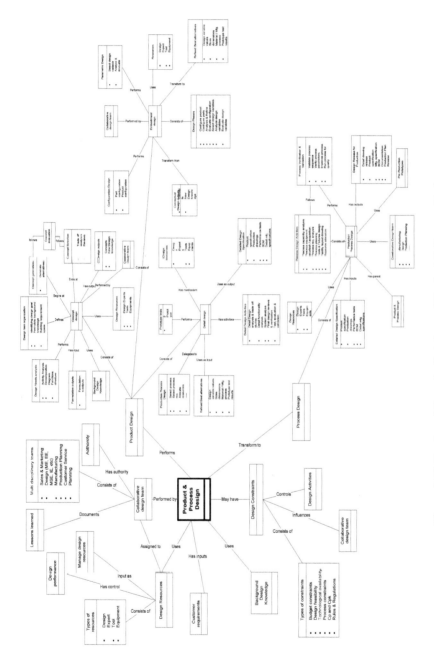

Figure 5.38 Top Level Domain Ontology of Product & Process Design

5.2.5 Publish Ontology

The sole purpose of publishing ontology is to share this knowledge with other ontology developers and/or with the design teams who need this to design their products and processes. Publishing ontology can be done in many ways. One of the easy ways to publish is to upload into a online repository, which is accessible to public. Protégé users upload their ontologies in their specific web site, which anybody can access. Some of the web addresses for uploading ontologies are http://www.ksl.stanford.edu/ software/ ontolingua, http://www.daml.org/ontologies, http://www.wines.com, etc.

This process reduces the data collection effort of an ontology developer significantly if, he/she develops similar domain of ontology or any site-specific ontology under this domain. Manufacturing enterprises can use this design ontology for their design needs.

CHAPTER 6

COMPANY DEMONSTRATION

6.1 Overview

The development of the Domain Knowledge Acquisition Process Methodology (DKAP) of product and process design was discussed in chapter four. Using the DKAP an ontology of the design domain was constructed & presented in chapter five. The next major task in this research work was to validate / demonstrate the methodology using real time companies. This chapter discusses the demonstration process using two companies around the Dallas Fort Worth (DFW) metroplex in the USA. Majority of the companies in this metroplex, who do the product and process design in house are somewhat related to defense industries. These industries were reluctant to provide design information because of the nature of their business. Two non-defense companies were selected to collect their design information and using those information, two different site-specific design ontology were constructed for verification of the DKAP methodology. Two companies were carefully selected from two different areas of design, one is good at product design and the other one is good at process design. In the following sections, brief descriptions of both companies as well as the implementation of methodology are discussed in detail.

6.2 Company Profiles

The two companies are referred to as Company A and Company B. Company A designs both products and processes and manufacture in-house. Company B does the process design only. Their descriptions are as follows.

6.2.1 Company A

Company A is an engineering company specializing in airport baggage handling systems, baggage security systems, cargo handling systems and parcel sortation systems. Since its establishment in 1972, Company A has grown to become a world leader in the field of airport baggage handling systems. It has over 33 years experience in airport systems design and installation with more than 449 projects in 38 countries around the world.

Company A operates from two main fully integrated manufacturing facilities in Dallas (TX), USA and Auckland, New Zealand. In addition to this it has subsidiary companies in Canada, Latin America and Australia. Its head office is located in Auckland, New Zealand where a highly skilled team of mechanical and electrical engineers, software designers, project managers and international sales managers support the company's technology and marketing programs.

In addition to state-of-the-art manufacturing facilities, Company A has an established and dedicated controls and software design department based at its head office in Auckland. The department has developed a comprehensive suite of system-compatible proprietary software designed to run on industry standard hardware providing low cost of ownership, reliability and future expandability. With a global network of representatives, strategic alliances and manufacturing facilities, Company A can quickly respond to customer requirements with internationally proven systems.

At the heart of its philosophy is a total commitment to quality, reliability and performance. Company A helps its customer operate efficiently under even the most demanding conditions by providing fast, accurate and secure movement of baggage. With the experience of over 350 installations worldwide, Company A offers a host of innovative solutions to keep the world moving.

Company A designs & manufactures engineering parts such as Flat pattern pier drive, Mounting Bracket, Tapered Roller, etc for baggage handling systems in the airport. The requirements of its customers are very specific and most of the time the products are standard. Customers specify their design needs very clearly in terms of part selections, product sizes and dimensions, material selections, etc. In this typical scenario, Company A designs some parts and produces in their plant and purchase standard parts from outside vendors. In case customers want some special parts, it has no option but to design, test, and manufacture that part in the plant. Even in case of special parts, customers specify the material, dimensions, and other functional requirements. This kind of specifications leaves no room for innovation in the design process but at the same time reduces the risk.

The design team of Company A is collaborative in nature. It consists of twelve experts from different disciplines. They are from management, production, design, outside vendor, customer service, and customers. Among all, production people and design people work very closely during the design process.

Basically Company A's design process produces detail drawings, part parameter values with relative arrangements, production process design, and sometime test results. Figure 6.1 shows an assembly drawing of some parts, which Company A needs to design and manufacture. Design team produces the detail drawings of the parts and assembly (see figure 6.1, figure 6.2, and figure 6.3) with the detail specification of design parameters, BOM, and process specification including types of processes and their sequences.

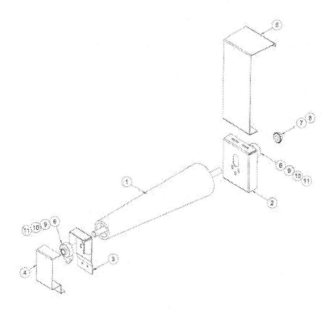

Figure 6.1. Exploded Assembly of 6^0 Tapered Roller with $30''$ Belt

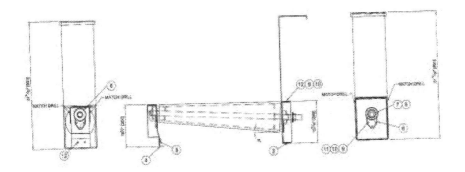

Figure 6.2. General Assembly of 6^0 Tapered Roller with $30''$ Belt

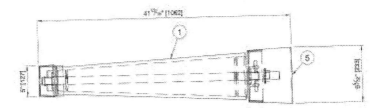

Figure 6.3. Top View of Tapered Roller

6.2.2 Company B

Company B is a world leader in providing geosynthetic-lining solutions. Its complete line of smooth and textured geomembranes combine long-term durability with strength, flexibility, and ultraviolet and chemical resistance to provide an impermeable barrier for a wide variety of applications. In addition, its geocomposite and anchor products allow us complement its remembrance solutions in the areas of waste disposal, fluid conveyance and containment, soil protection and stabilization, and gas recovery systems. Applications for its products include solid and hazardous waste landfills, landfill closures, secondary containment, ponds, reservoirs, fish hatcheries, irrigation canals, wastewater treatment plants, and industrial waste lagoons, etc.

Company B was founded in 1976. It is part of a group of privately held corporations, which produces several product lines, all within the polyethylene family. The group of companies comprises the world's largest producer of polyethylene construction film, the highest quality trash bag manufacturer in the United States, and a leading supplier of geomembrane liners. Company B is also the most technologically advanced and largest recycler and compounder of polyethylene in the world. Company B is dedicated to providing the highest quality polyethylene products and its staff of chemical and process engineers utilize all available modern technology. The company's design team is responsible for several technological breakthroughs in blown-film extrusion, bag making and recycling. In order to support its research and quality control efforts, it maintains one of the most sophisticated on-site laboratories in the plastic industry.

Research and development is one of the core competencies for Company B. It has a good record of innovating new products and standards. Due to the nature of polyethylene production, the product design process is very brief. Product design consists of identifying design needs, selection of ingredients, identifying proportion of ingredients, in some cases drawing specifications, and testing. Depending on the products, it has to do a detail process design before actual fabrication of products.

Company B produces the following products in its plant and uses approximately eight different production processes.

- Trash bags

- Construction film

- Stretch film

- Painter's plastic

- Flame retardant

- Carpet film

- Geomembrane

Grades of polyethylene such as low-density polyethylene (LDPE), linear low-density polyethylene (LLDPE), high-density polyethylene (HDPE) and ethylene-vinyl acetate copolymers (EVA) are available to meet the needs of various fabrication processes. Different processes are blown and cast film, blow molding, injection molding, extrusion coating, and laminating. The ease of processing and high purity makes polyethylene an extremely attractive material capable of competing quite favorably in a number of demanding applications. A detail design specifications of a sample polyethylene product could be as follows.

The raw material shall be made of polyethylene resins manufactured in the United States. Carbon black shall be added to the resin if the resin is not precompounded for ultra-violet resistance. Blow molding will be used to fabricate the product and testing must conform the values provided in the Table 6.1. The list of properties varies from product to product and customer needs.

Table 6.1. Seam Properties of a Sample Polyethylene Product

Seam Property	20 mil	30 mil	40 mil	60 mil	80 mil
Shear Strength, lb/in	26	40	53	79	105
Shear Elongation, %	50	50	50	50	50
Peel strength, lb/in	22	34	44	66	88
Peel separation, %	25	25	25	25	25

A preproduction prototype is necessary in this situation to conform the functionality of product. The final sample product must meet the following values for its specific uses.

Density: ASTM D1505 $\geq$ 0.940 g/cc

Melt Index: ASTM D1238 $\leq$ 0.4 g/10 minutes

Carbon Black Content: ASTM D1603 2% - 3%

Tensile Strength at Yield: ASTM D6693 2,100 lb/in2

Dimensions: As shown on the drawing on file

Weight: 0.45 lb/ft

Length: 5 ft and 10 ft

6.3 Development of Questionnaire & Implementation Guideline

To make the data collection process smooth and efficient, a questionnaire was developed along with an interview kit. Interview Kit is document for conducting formal interviews with the participating companies using questionnaires or other supporting materials. A standard interview kit contains the following.

- Introduction

- Research Overview

- Problem Statement

- Research Objective

- Purpose of Interview

- Interview Methodology

- Selection of Interviewees

- Topics of Interview

- Types of Information sought

- Data Collection Method

- Detail Questionnaire & Follow-up

Before collecting any information, it is necessary to mention about the procedure of interview methodology. This gives an idea to the interviewee and facilitates better communication. It was also mentioned about the time and resources needed from the participating companies and what kind of & how detail information were looking for. In this effort, interview methodology mentions about asking access to design team members, mid level managers, shop floor level production managers, and workers.   The initial plan was to interview approximately 10 people from each participating company. The interview process consisted of two sessions with the collaborative design team members and each session lasted over two hours depending on the interviewee. A questionnaire was developed carefully to collect sufficient design information about the company necessary to build the ontology. The interview was in-depth with organization personnel. The interview was guided by questionnaire. Sometimes, the question was explained during the interview in a way that was comprehensible to the interviewee. The same questionnaire was used for both the companies to ensure consistency in the data to be collected. Both the interview kit and the questionnaire are attached in appendix B.1.

There is an implementation guideline that is attached in the "Appendix B.2" that guides the implementation of each and every step given in the revised methodology.

6.4 Refined Methodology

The interview process exposed some of the issues in the methodology. The suggestions from the industry experts are taken into consideration, analyzed for practicality and fitted into the final methodology. In addition, during the course of the interview, there were some discrepancies, which are clearly evident, are also taken into account in the final revised methodology. This section talks about the final methodology that the manufacturing enterprise could use to build domain or site-specific ontology of their design processes.

6.4.1 Determine the Domain and Scope

Product and process design is a very complex, innovative, and iterative process. Though, it follows structured steps of processes, it varies from product to product. In case of high-end electronic products, conceptual design and testing of products are more important than that of a matured product such as automobile. Product and process design includes industrial design, engineering design, and production design (Rudolph, 2005). It is a part of product development life cycle and product development is again a part of product life cycle. A product life cycle consists of the following steps. Figure 6 4 shows these steps with further decomposition of product development.

i.   Product development

ii. Production

iii. Distribution

iv. Service

v.   Disposal

Product development is decomposed into sales & marketing and product and process design. As mentioned earlier that product and process design consists of industrial design, production design, and engineering design.

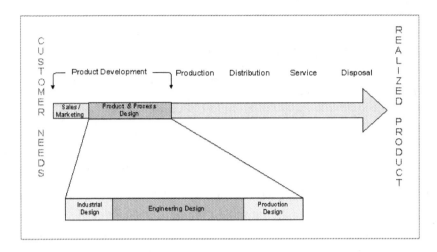

Figure 6.4. Product Life Cycle Phases (Modified from Rudolph, 2005)

Product and process design itself is a huge domain. It is almost always nice to capture all available entities and relations in the domain but sometime impossible and long-term project. Neither company A nor company B performs original product and process design. Most of the time company B performs variant and/or selection design and sometime configuration design. Company B performs only selection design along with process design. Different design processes are shown in figure 6.5.

136

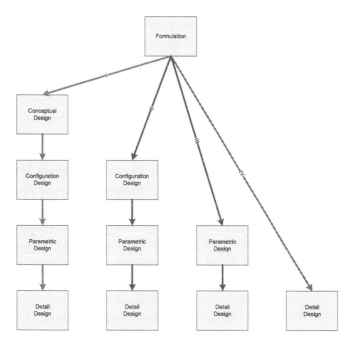

Figure 6.5. Types of Design processes (I: Original design; II: Configuration design; III: Variant design; IV: Selection design)

The scope of the domain was limited to specific part of product and process design, which each company performs with respect to their collaborative design team. This type of ontology is called site specific ontology and very useful

6.4.2 Check Availability of Existing Ontologies & Reuse

It is almost always worth considering what someone else has done and checking refinement and extends existing sources for design domain and task. There is no valid reason to expend resources to build an ontology, which is already available. In some cases, a similar kind of ontology can be derived from the available one. Reusing existing ontologies may be a requirement if the system needs to interact with other applications that have already committed to particular ontologies or controlled vocabularies (Natalya & Mc Guinness, 2001). For example, a knowledge base of enterprise integration ontology may already exist. If one can import this knowledge base and the ontology on which it is based, he/she will have not only the classification of enterprises but also the first pass at the classification of integration characteristics used to distinguish and describe the terms.

The Product and process design domain ontology, which was developed in chapter five, was used to develop the site-specific ontology for company A and company B. This reuse not only drastically reduces the time and effort of ontology building but also ensures conformity.

6.4.3 Organize the Project

This activity will set different task to be performed to build the new ontology. Some of the tasks are to form Development Team, break down the tasks, assign team members to specific tasks, etc.

An important initial step in developing an ontology description is the formation of a development team. Each member of the team will perform a particular role in the development effort. Individuals who are involved in the modeling may each fulfill several roles, but each role is dealt with distinctly and should be clearly separated in the minds of the participants. The following are the sample roles assumed by the ontology development project personnel:

i.   Project Leader: This administrative role is responsible for overseeing and guiding the entire ontology development effort.

ii.   Analyst/Knowledge Engineer: Personnel with ontology development expertise who will be the primary developers of the ontology description fill this technical role.

iii.   Domain Expert: This role characterizes the primary sources of knowledge from the application domain of interest. Persons filling this role will provide insights about the characteristics of the application domain that are needed for extracting the underlying ontological knowledge.

iv.   Team Members: All persons involved with the ontology description project.

A Work Breakdown Structure (WBS) was constructed and team members are assigned against each individual task to ensure the progress of the ontology building effort. In this particular ontology, I was responsible to do everything with the help of industry domain experts.

6.4.4 Data Collection and Analysis

One of the major and time-consuming tasks of this effort was to collect data related to the domain of interest. An extensive literature reviews were conducted to identify product and process design related terminologies, statements, relative importance of design steps, new principles in product and process design, and so on. All terminologies identified in chapter five are used as term pool for this site-specific ontology. The term pool was refined once design experts were interviewed.

Survey questionnaires are designed to collect design information from the design experts. In this questionnaire, design experts were asked to identify their design steps, list of activities under each steps, relative importance of each design activity in the design process, and how these activities are related to each other. Table 6.1 was used to rank the importance each design activities and a matrix shown in figure 6.6 was used to determine the interrelationship among design activities. A sample questionnaire is presented in the Appendix B1.

Table 6.2. Ranking Design Activities According To Their Importance

List of activities	Rank (0 – 9)
I.	
II.	
III.	
IV.	
V.	
VI.	
VII.	
VIII.	
IX.	
X.	
XI.	
XII.	
XIII.	
XIV.	
XV.	
XVI.	
XVII.	
XVIII.	
XIX.	
XX.	
XXI.	
XXII.	
XXIII.	
XXIV.	
XXV.	

Three types of relationships are identified among design activities. Following symbols represented those relationships.

"xx " for strong relationship

"x" for moderate relationship and

"o" for no relationship

List of Design Activities	Activity # 1	Activity # 2	Activity # 3	Activity # 4	Activity # 5	Activity # 6	Activity # 7	Activity # 8	Activity # 9	Activity # 10	Activity # 11	Activity # 12	Activity # 13	Activity # 14	Activity # 15	Activity # 16	Activity # 17	Activity # 18	Activity # 19	Activity # 20
Activity # 1																				
Activity # 2																				
Activity # 3																				
Activity # 4																				
Activity # 5																				
Activity # 6																				
Activity # 7																				
Activity # 8																				
Activity # 9																				
Activity # 10																				
Activity # 11																				
Activity # 12																				
Activity # 13																				
Activity # 14																				
Activity # 15																				
Activity # 16																				
Activity # 17																				
Activity # 18																				
Activity # 19																				
Activity # 20																				

Figure 6.6. Interrelationship Matrix

More than two hundred terminologies are identified from this data collection effort. A partial snapshot of this data is shown in the following figure.

Anticipatory failure
determination
Assembly drawings
Axiomatic design
Backtracking
Benchmark
Bending
Beta prototype
Beta testing
Bill of materials
Blanking
Blow molding
Boring
Brazing
Built-in-self-test
Bulk deformation
Casting
Clearance fit
Coefficient of thermal
expansion
Collaborative design
team
Company requirements
Component
decomposition
Compression molding
Computer-aided design
Computer-aided
engineering
Configuration design
Configuration
requirements sketch
Consensus decision
Contingency design
Correlation ratings
matrix
Customer requirements
Decision-making
process
Design analysis
Design compatibility

Design for testability/
inspect ability
Design for
serviceability
Design for cost
Design for lifecycle
cost
Design for performance
Design for reliability
Design for safety
Design for assembly
Design for disassembly
Design for adjustability
Design for assembly
Design for close fit
Design for manufacture
Design for the
environment
Design for the extreme
Design for x
Design goal
Design guidelines
Design method
Design need
Design phase
Design problems
Design process
Design requirements
Design reviews
Design validation
Design variable
Design verification
Design-project report
Detail design
Detail drawings
Diagram
Die casting
Drawing
Drilling
Early supplier
involvement

Embossing
Engineering analysis
Engineering change
Engineering design
Engineering design
specification
Equipment
Evaluation
Evaluation criteria
Extrusion
Facing
Fail-safe design
principle
Failure analysis
Feasible design
Field-testing
Finishing
Fixture
Forging
Forming
Formulation
Function decomposition
Functional test
Grinding
Hazard analysis
Highly accelerated life
test
Honing
Human factors
Industrial design
Injection molding
Insertion
Investment casting
Lapping
Machine
Machine tool
Machining
Manufacturing Design
Mechanical press
Milling
Mistake-Proofing

Figure 6.7. Partial Term Pool of Site Specific Ontology

6.4.5 Initial Design Ontology

Once term pool was compiled, data analysis begins. In this effort data analysis was conducted in accordance with IDEF5 methodology. A list of proto-kinds and a list of proto-relations are constructed using the top down approach, mentioned in chapter four. Using IDEF5 analogies and consistency check, a final list of kinds/ entities and relations are derived. In the following section, a detail description of all identified entities/kinds and relations are presented.

6.4.5.1 Entities of Design Ontology

The elements described in this section are the entities of the site-specific ontology of products and processes design for company A & Company B. Identification of these design entities was done using a well-structured method of IDEF5 modeling tool. Two major types of entities are identified in this book. They are design activity and design object. Since the domain of interest is products and process design, which itself an activity, majority of entities in this effort fall under this category. Design activities are any process of transforming state of the environment that happens over time. Other type is considered as resources, mechanisms, controls, inputs, and outputs of the design processes. An object is defined as a tangible or conceptual entity in the design domain. Object is further classified into two sub categories; they are Actor and Products. Actor could be a design expert, collaborative design team, or other entity capable of actively participating in an activity. Actor is classified into three sub categories. The taxonomy of the design entity is shown in figure 6.8. Both company A and Company B support this structure for their design entity.

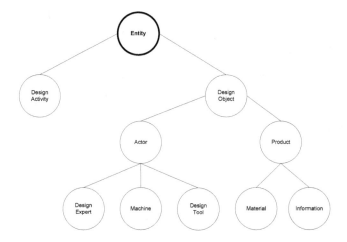

Figure 6.8. Taxonomy of Design Entity

A design expert is defined as a human being entity that performs activities to achieve design goals. Its ability to actively participate in the planning process is what separates a Design Expert from a Machine. A machine is a non-human entity, which has the capacity to carry out design functions. Design Tool is also known as a non-human entity, which has the capacity to carry out design functions. Product category of entities in the design domain is considered as any passive object that can be used, created, or modified by an activity in the design domain. These are typically the inputs and outputs of a design process. Product entity has Material and Information type of design entities. Material is an object that can be used, created, or modified or consumed by an activity in the design domain. (e.g. Raw materials, Subassemblies). Information is defined as an

144

object that can be used, created, or modified by an activity and whose essence is to act as information in the design domain.

Each entity has list of attributes and defining properties, which makes them distinguishable. Table 6.3 shows detail description of a design activity type entity called "Process Design". IDEF5 supports this form of representation, which is hierarchically structured.

Table 6.3. Process Design Entity

1.1 PROCESS DESIGN	
Defining and designing a manufacturing process to accommodate the specific requirements of a given product while, meeting process quality and cost objectives.	
Attribute	**Description**
Name	Process design
Number	A unique qualifier for the activity
Parent	Product & Process Design
Documentation	User defined description of the process design
User-Attributes	User defined attributes for process design
Components	ACTIVITIES that make up this Activity. Such as identifying processes, sequencing processes, test and evaluation, validation and verification, etc.
Authorizer	The agent responsible for this activity, usually the collaborative design team is considered the process owner and has authority over process design.
Objects	The objects (non resource) used by this Activity, Objects in this case are material and information.
Resources	The collaborative design team and machine that performs the Activity
Sub kinds	Production process planning, Verify and validate process

Another example of an object category entity called "Design expert" is shown in Table 6.3. This type of representation scheme gives a clear description of entity.

Table 6.4. Design Expert Entity

2.1.1   DESIGN EXPERT
A human being entity that performs activities to achieve design goals. Its ability to actively participate in the planning process is what separates a DESIGN EXPERT from a MACHINE

Attribute	Description
Name	Design Expert 's name
Number	A unique qualifier for the Design Expert
Parent	Actor
Documentation	User defined description of the Design Expert
User-Attributes	User defined attributes for a Design Expert
Performs	The ACTIVITIES, which this DESIGN EXPERT is performing.
Sub kinds	Component of kind

In appendix A2, all entities of the design domain, identified in this effort are presented in the order of hierarchical relationship. All entities except conceptual design entities are relevant to site ontology for company A. All entities except industrial, conceptual, configuration and parametric design entities are relevant to site ontology for company B. In the following sections, some of these identified entities are further discussed and presented in a visual form, which is supported by IDEF5 representation scheme. In some cases a modification was made to draw figures to accommodate additional information. Entities in the domain do not mean anything without the relationships among each other. In the immediate following section, a detail description of such relations in the design domain is presented.

6.4.5.2 Relations of Design Ontology

Relationship is the way to connect, associate, or relate two or more entities with each other (Uschold *et al.*, 1995). It defines the interaction of entities. In products and processes design, this relation describes how the entities are associated and how the overall design function is performed. Figure 6.9 shows different relations with Parametric Design entity and with other design entities.

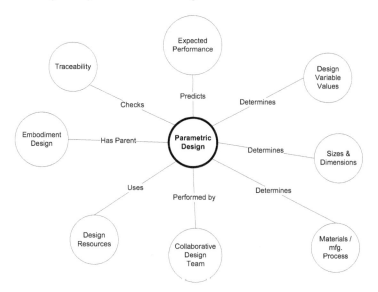

Figure 6.9. Relations of Parametric Design Entity with Other Entities

Each of these relations shown in the figure 6.9 is described individually using the IDEF5 format. Table 6.4 shows an example relationship specification for relation "Performed by" of figure 6.9.

Table 6.5. Example Relationship Specifications

1. PERFORMED BY		
Entity 1	**Entity 2**	**Definition**
Parametric Design	Collaborative Design Team	A person, group of person, or machine performs the activity "Parametric Design"

IDEF5 allows the differentiation of relationships according to the importance in the design processes. Two types of relationships are first order or primary relationships and second order or secondary relationships. Primary relationships are represented with an arrow in the head and secondary relations are presented with an arrow in the tail. In figure 6.10, component of systems design for a matured chemical product is described, where manufacturing process design, quality design, testability design, safety design, and disposal design have first order relationships with the systems design and the rest have secondary relationships with the systems design.

This type of relationships changes from industry to industry, product to product, and company to company. For example, electronic industry requires more importance on product development and testing whereas a pharmaceuticals company emphasizes on safety and quality of product.

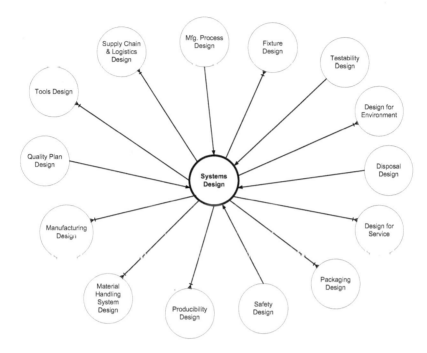

Figure 6.10. Components of Systems Design for a Matured Chemical Product

6.4.5.3 Descriptions of Some Design Entities

This section describes the representation scheme for some entities of site-specific ontologies for company A and company B. It provides an elaborate description of entities, its components, and relations with other entities. As mentioned earlier, that product and process design is a series of sequential activities, which transform customer's requirements into production release with detail engineering specification and detail process specifications.

6.4.5.3.1 Industrial Design & Formulation

These are processes of transforming customer/design requirements into design requirements using design resources and satisfying design constraints. This activity ensures feasibility of the design. In case of company A, this activity is known as requirements translation, which is the most important design activity for them. A detail description is shown in figure 6.11.

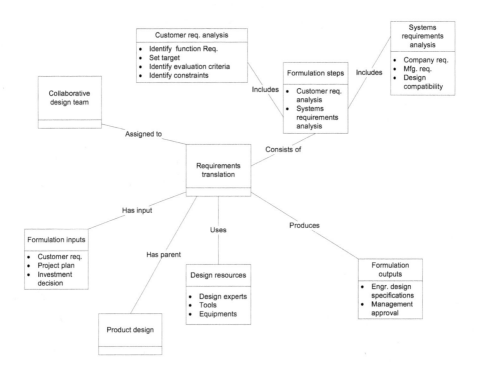

Figure 6.11. Requirements Translation for Company A

6.4.5.3.2 Embodiment Design

It is a process of selecting detail specifications of product and it's part with the consideration of various aspect of design such as safety, manufacturability, assembly, environment, etc. It has two sub kinds/ entities; configuration design and parametric design. Both configuration design and parametric design has their own sub kinds. A detail description is shown in figure 6.12. In some cases, company A does not require this function and company B never does this function to do their business.

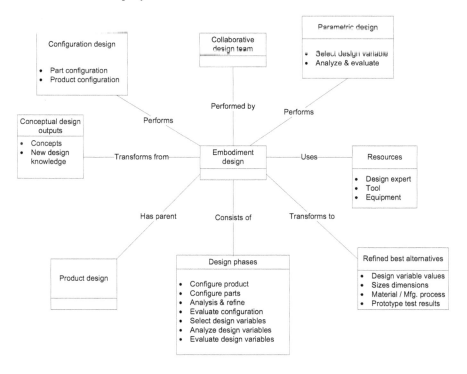

Figure 6.12. Embodiment Design Process

6.4.5.3.3 Configuration Design

Configuration Design is a process of selection and arrangement of features on a part; or the selection and arrangement of components on a product; a phase of design when geometric features are arranged and connected on a part, or standard components or types are selected for the architecture. A detail description is shown in figure 6.13.

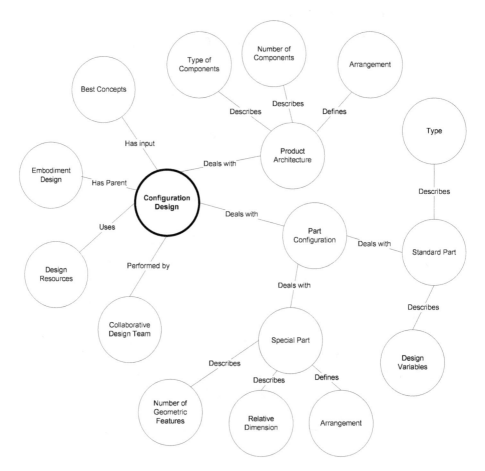

Figure 6.13. Configuration Design Process

152

6.4.5.3.4 Parametric Design

It is a phase of design that determines specific values for the design variables. A detail description is shown in figure 6.14.

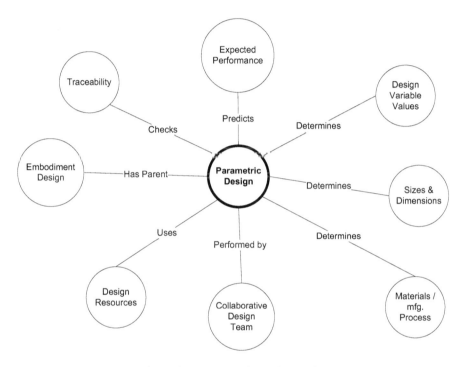

Figure 6.14. Parametric Design Entity

6.4.5.3.5 Detail Design

Detail Design is a phase of design that results in the preparation of a package of information that includes drawings and specifications sufficient to manufacture a product. A detail description is shown in figure 6.15. In the detail design, company A establishes various cost of the design, perform EVMS, and reviews design in various

phases.

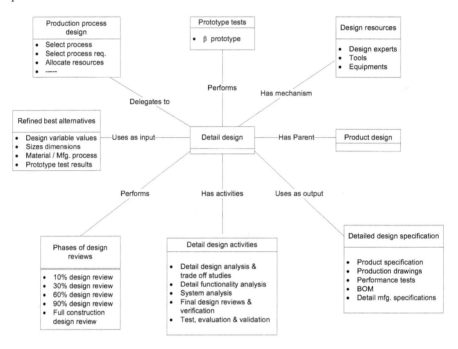

Figure 6.15. Detail Design Schematic for Company A

6.4.5.3.6 Detail Design Analysis

It is a phase of design that predicts or simulates performance of each alternative,

reiterating to assure that all the candidates feasible. A detail description is shown in

figure 6.16.

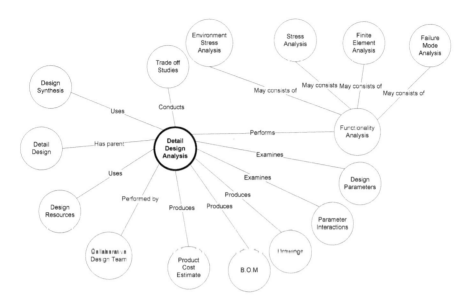

Figure 6.16. Detail Design Analysis Process

6.4.5.3.7 Design Reviews & Verification

These are phases of design that identifies technical performance, risks, improvement of performance & process, and verifies design parameters. A detail description is shown in figure 6.17. Company A perform the design reviews in several stages like 10%, 30%, 60%, 90%, and full construction to ensure that cost and schedule are on right track.

155

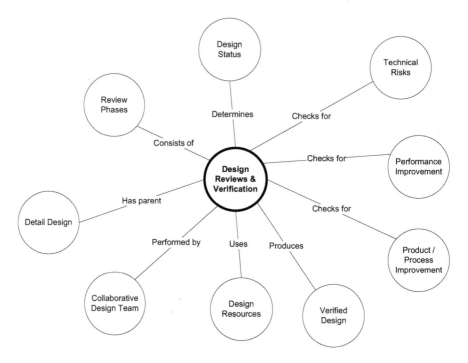

Figure 6.17. Design Reviews & Verification

6.4.5.3.8 Test & Validation

Test & Validation are processes of evaluating each design specifications and validation through extensive testing and simulation. Company B does not perform all tests because the nature of the product it produces. A detail description of test & validation for company B is shown in figure 6.18.

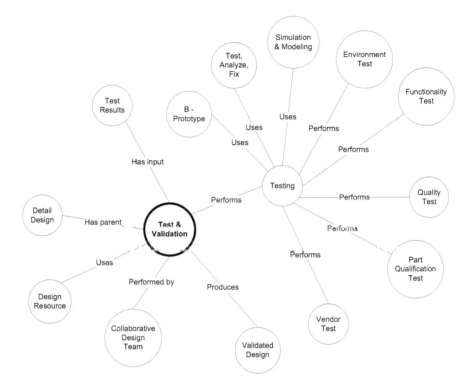

Figure 6.18. Test & Validation in Detail Design Process

6.4.5.3.9 Detail Systems Design

It is a phase of design that determines the required systems specifications for the product to be produced. A detail description is shown in figure 6.19.

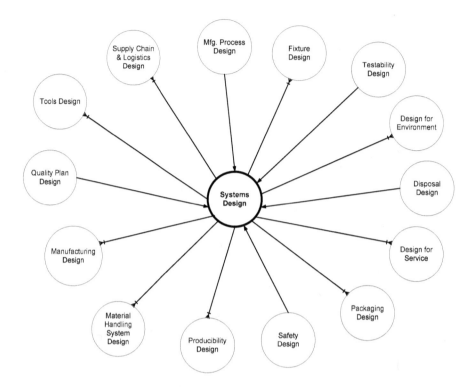

Figure 6.19. Detail Systems Design

6.4.5.3.10 Production Process Design

Defining and designing a manufacturing process to accommodate the specific requirements of a given product while, meeting process quality and cost objectives. Company B considers this activity as the single most important one because of the nature of the product it produces. A detail description is shown in figure 6.20.

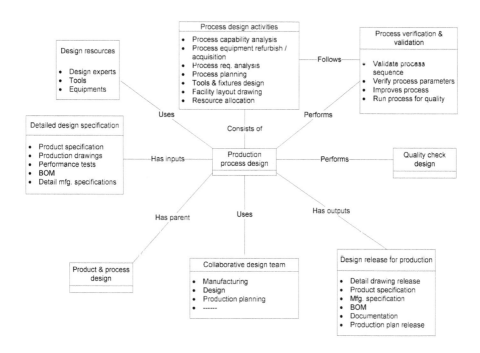

Figure 6.20. Production Process Design Entity for Company B

6.4.5.3.11 Process Design Analysis

It includes process capability analysis, process requirements analysis, and production process planning, which is a process of planning and determining process specifications to produce the designed products. A detail description is shown in figure 6.21.

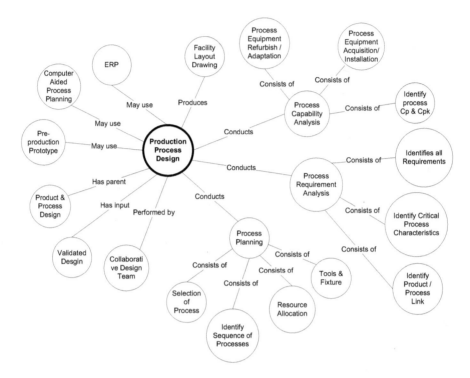

Figure 6.21. Process Design Analysis

6.4.5.3.12 Design Resources

These are mechanisms or inputs in the design process. They can be design experts, equipments, or design tools. A detail description is shown in figure 6.22 and component of design resources is shown in figure 6.23.

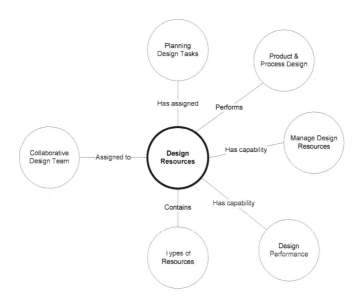

Figure 6.22. Design Resources

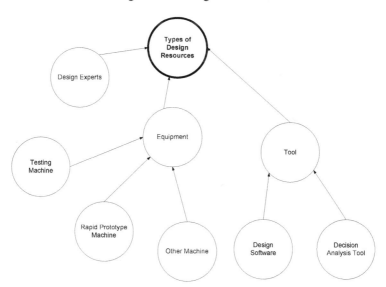

Figure 6.23. Types of Design Resources

161

6.4.5.3.13 Collaborative Design Team

A multidisciplinary design team consisted of people including management, designers, product support, vendors, and customers. The key objective is to improve communication in the design process. A detail description is shown in figure 6.24.

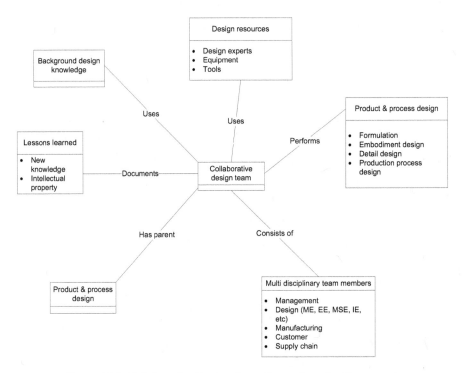

Figure 6.24. Collaborative Design Team Formations for Company A

6.4.5.3.14 Design Constraints

These are obstacles to design products and processes. They can be financial, technical or administrative in nature. A detail description is shown in figure 6.25.

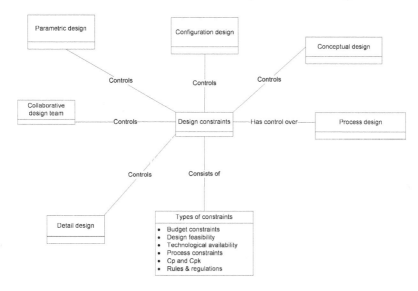

Figure 6.25. Design Constraints

6.4.5.3.15 Hierarchy & Sequence in the Design Process

Entities in the design domain are hierarchically structured. Complete hierarchy can be found in Appendix A1. For company A, actual Product & process design starts at configuration design and ends at production process design. For company B, actual Product & process design starts at detail design and ends at production process design. Each steps has several sub steps. All these steps and sub steps are sequential and dependent on the predecessor steps. Detail descriptions of each design steps are shown in figure 6.26 through figure 6.30.

163

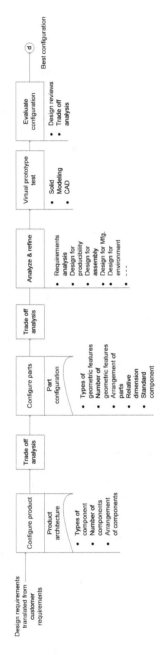

Figure 6.26. Configuration Design Process for Company A

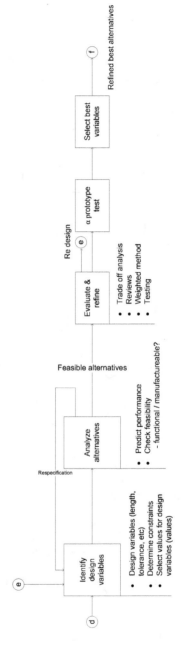

Figure 6.27. Parametric Design Process for Company A

164

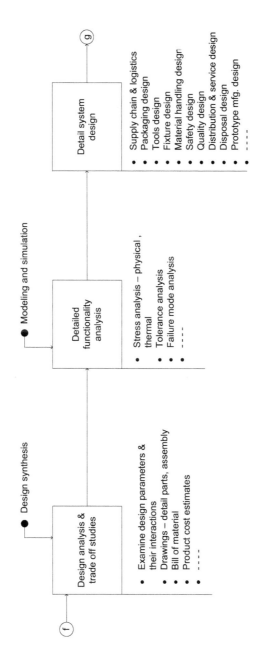

Figure 6.28. Detail Design Process for Company A

165

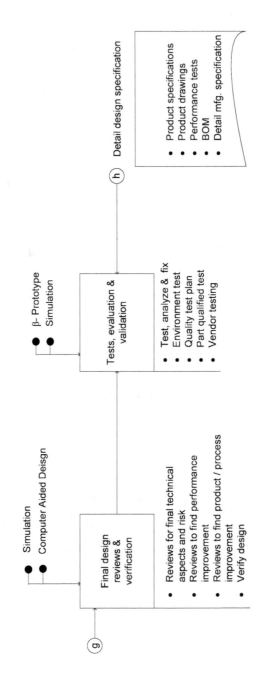

Figure 6.29. Detail Design Process Final Stages for Company A

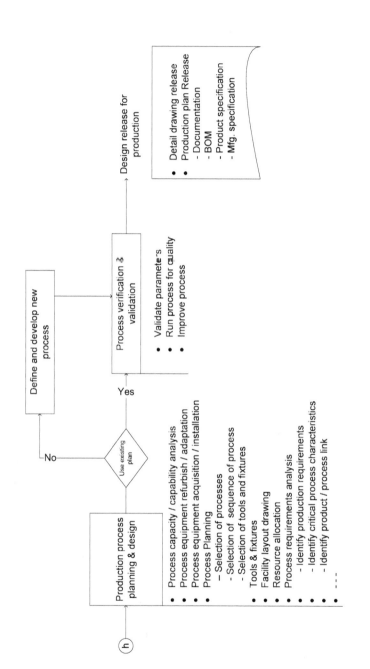

Figure 6.30. Production Process Design for Company B

6.4.6 Refine & Validate Ontology

The entities, properties, attributes, and relations defined and developed in the previous sections are refined and validated using the IDEF5 methodology. This activity will affirm the validity of the design ontology to the products and processes design domain. Typically this step will identify the need to reiterate some of the steps accomplished earlier. It resolves the contradictory concepts or definition of design terms in the ontology. The outputs of the previous steps are definitely pieces of ontology. As mentioned earlier, ontology has three basic things. They are terminologies, entities, and relations. For example, terminologies are similar to dictionary, which define the terms of the domain. Entities/kind and sub kinds are the building block of an ontology. Each entity has list of attributes and defining properties, which are unique and separate from other entities. Relations are also very important building block within the ontology. It provides ontology semantics, which is absent in the dictionary.

Validation of this site ontology was done by comparing and contrasting against the domain ontology already developed in chapter five. The consistency check was conducted in the following section.

As part of the refinement and validation process, a complete site specific ontology model for company A & company B of their products and processes design are shown in figure 6.31 and figure 6.32 respectively.

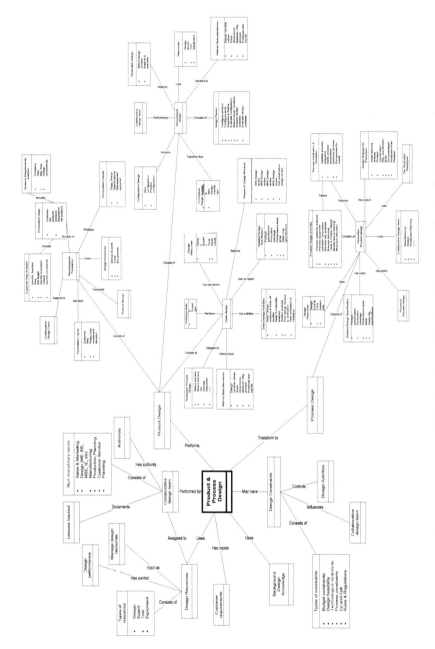

Figure 6.31. Top Level Site Ontology of Products & Processes Design for Company A

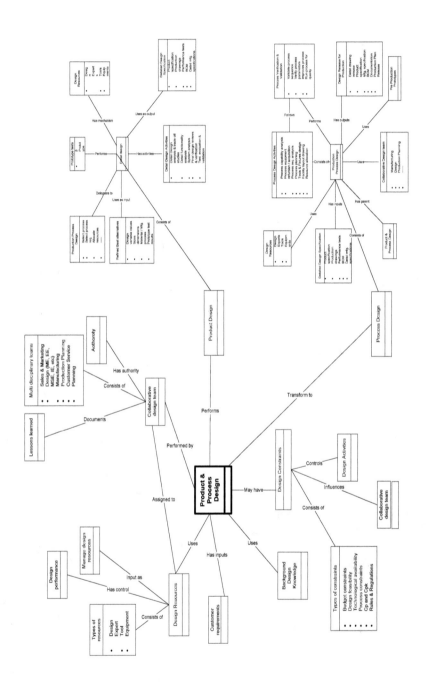

Figure 6.32. Top Level Site Ontology of Products & Processes Design for Company B

6.4.7 Check Consistency & Accuracy of Ontology

To ensure accuracy and consistency of information in the ontology, a consistency matrix, which was developed in chapter four, was used in this section. This methodology is straightforward and easy to use for manufacturing enterprises. It has the following steps.

1. Identify the design activities with relative importance of the domain of interest

2. Prioritize design activities of the domain

3. Check captured design activities against the domain design activities

4. Calculate consistency index

Design activities are identified from the domain ontology already developed in chapter five. These activities are listed in table 6.5. The relative importance was identified from the evaluation of design experts using the questionnaire.

Table 6.6. Top Level Design Activities

Industrial Design	Identify design variables
Customer requirements analysis	Analyze alternatives
Review requirements	Evaluate & refine variables
Feasibility analysis	Prototype test
Identify design needs	Detail Design
Conceptual Design	Design analysis & trade off studies
Analysis of design needs	Functionality analysis
Identify design tasks	System design
Generate concepts	Design reviews & verification
Evaluate concepts	Test & evaluation
Select concepts	Design validation

Table 6.6 - continued

Embodiment Design	Production Process Design
Configure product	Process planning
Configure part	Process requirements analysis
Trade off analysis	Process capability analysis
Analyze & refine configuration	Process selection
Prototype test	Process parameter selection
Evaluate configuration	Process verification & validation

Present the result graphically by plotting the percentage of total cumulative value on the Y-axis and the item number in X-axis. A typical Pareto curve for a design situation is shown in figure 6.33.

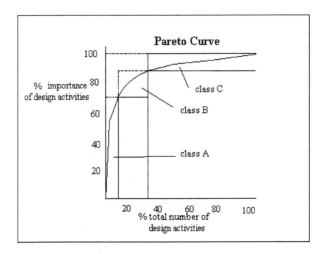

Figure 6.33. Pareto Curve of a Design Situation

From the above analysis only five activities are considered as class A activities, which accounts for about 70% importance in the design process. Eight activities are considered as class B activities, which accounts for about 15% importance in the design process. The rest eighteen activities are considered as class C activities. All design activities, identified for the site-specific ontology, are then checked against these three classes of activities. Figure 6.34 shows checking of design activities for company A.

Consistency Check Matrix for Company A — mapping of Captured Design Activities in the Design Ontology (rows) against Design Activities classified as Type A, Type B, and Type C (columns). Each check mark (✓) indicates the matched design activity.

Captured Design Activities in the Design Ontology	Matched Design Activity (column)	Type
Conceptual		
Customer requirements analysis	Requirements analysis	A
Review requirements	Review requirements	C
Identify design needs	Identify design needs	C
(—)	Evaluate concepts	A
Functional Analysis	—	
Embodiment		
Configure product	Configure product	C
Configure part	Configure part	C
Trade off analysis	Trade off analysis	B
Analyze & refine configuration	Analyze & refine configuration	C
Evaluate configuration	Evaluate configuration	C
Identify design variables	Identify design variables	C
Analyze alternatives	Analyze alternatives	C
Evaluate & refine variables	Evaluate & refine variables	C
Detail		
Design analysis & trade off studies	Design analysis & trade off studies	C
Functionality analysis	Functionality analysis	A
Design reviews & verification	Design reviews & verification	B
Test & evaluation	Test & Evaluations	A
Design validation	Design validation	B
Process		
Process planning	Process planning	C
Process requirements analysis	Process requirements analysis	B
Process capability analysis	Process capability analysis	C
Process selection	Process selection	C
Process parameter selection	Process parameter selection	A
Process verification & validation	Process Verification & validation	B

Row Score	Type A	Type B	Type C
	5	5	13

Figure 6.34. Consistency Check Matrix for Company A

173

Calculation of consistency index: This index provides a clear picture of the accuracy and consistency of design information, which is captured to construct the ontology. Index calculation was done using the following steps.

a.   Normalize the raw score- From the consistency cheek matrix in figure 6.34, three different raw scores are found; they are raw score for type A activities ($R_A = 5$), raw score for type B activities ($R_B = 5$), and raw score for type C activities ($R_C = 13$). Normalization of these scores was done using the following equations.

$$N_A = \frac{R_A}{M_A} X100$$; where N_A represents the normalized score for type A activities and

M_A represents the maximum raw score for type A activities. In this particular case M_A =5, there fore $N_A = 5/5*100 = 100$

$$N_B = \frac{R_B}{M_B} X100$$; where N_B represents the normalized score for type B activities and

M_B represents the maximum raw score for type B activities. In this particular case M_B =8, there fore $N_B = 5/8*100 = 62.5$

$$N_C = \frac{R_C}{M_C} X100$$; where N_C represents the normalized score for type C activities and

M_C represents the maximum raw score for type C activities. In this particular case M_C =18, there fore $N_C = 13/18*100 = 72.2$

b.   Calculate the index- Pareto analysis is based on Pareto principle, which says that typically, type A accounts the first 20% of activities in the list will account for approximately 70% of cumulative importance. The next type B accounts 30% of activities, will, typically, account for a further 15% of cumulative importance. These can be subject to less precise control methods. The last type C accounts for the rest 50% of (low importance) activities then account for a mere 15% of importance and can be controlled with a simple system. From this basic principle, the relative weight of each type of activities can be presented as W_A =0.7, W_B =0.15, and W_A =0.15, where W_A, W_B, and W_C are the relative weight of type A, type B, and type C activities respectively.

$$CI = W_A N_A + W_B N_B + W_C N_C$$

or $CI = W_A \left. R_A \middle/ M_A \right. X100 + W_B \left. R_B \middle/ M_B \right. X100 + W_C \left. R_C \middle/ M_C \right. X100$

$CI = 0.7*100+0.15*62.5+0.15*72.2 = 90.2$

The value of consistency index for company A is 90.2, which implies that the site specific ontology developed for company A is accurate and consistent to the domain ontology of product and process design. If the value of this index is low for design ontology, there is question of validity of such ontology model.

6.4.8 Publish Ontology

The sole purpose of publishing ontology is to share this knowledge with other ontology developers and/or with the design teams who need this to design their products and processes. Publishing ontology can be done in many ways. One of the easy ways to publish is to upload into an online repository, which is accessible to public. This process reduces the data collection effort of an ontology developer significantly if, he/she develops similar domain of ontology or any site-specific ontology under this domain. Manufacturing enterprises can use this design ontology for their design needs. Both company A & company B realized this need and glad to use the domain ontology, which was developed in chapter five. But both have reservation about publishing their own site specific ontology for the fear of revealing their secrets. They have recommended incorporating this information into domain ontology and publishing the updated domain ontology for others to use.

6.5 Summary

This book discusses several contributions to the engineering literature. The first contribution is a comprehensive methodology to construct design ontology for manufacturing enterprises. This methodology is tailored to capture design knowledge, build ontology, and ensures accuracy and consistency of captured design information. In addition, the methodology has the following unique characteristics:

- It introduces the notion of reusing ontologies within the similar domain to reduce ontology building effort. Unfortunately there is no design ontology available for reuse, but one can surely use the design ontologies, which was developed in chapter five.

- This methodology has mechanism to ensure the accuracy and consistency of captured design knowledge. This mechanism is vital for manufacturing enterprises that perform product and process design.

- It is able to incorporate lessons learned during the design process and facilitates reuse of design information. This makes the design ontology continuously up-to-date.

- It publishes the design ontology for others to use in developing their own ontologies or for their design needs.

This book discusses about cross consistency matrix, which has a mechanism to check accuracy and consistency of captured information against domain ontology. This type of cross checking is vital for design ontology and is not in place in any ontology building effort. This matrix is easy to construct and unique in the design domain.

It describes a domain ontology model for generic product and process design which, can serve as a source of design knowledge for enterprise engineering. A direct benefit of such ontology is the enhanced ability to perform product and process design. The ontology, built on IDEF5 language, will allow for common understanding and unambiguous meaning for each design entity and relationship. The result is that different design experts studying the model will have a common basis of communication and understanding. A common understanding is critical in product and process design, especially in collaborative design efforts, which will undoubtedly involve many individuals from different functions, fields and disciplines. The entities defined in the ontology, especially the Design Activity entity described in chapter five, can be used as parts of the business process templates. The Design Activity entity defines the generic properties, characteristics and relationships which are common for most activities. This domain ontology can be helpful for both design & research community in following ways:

- Any design team can use this domain ontology for their own product and process design or can get the guideless for their design process.

- Ontology use may reduce the design lead-time & design errors and hence reduce cost of design.

- Facilitates better communication among top management, other departments, suppliers, external partners, sub-contractors, etc.

- Activity model and process model can be derived from an ontology model.

- Ontology developer can use this domain ontology to build similar domain ontology or site-specific ontology in the area of product and process design.

Appendix B presented an implementation guideline. This implementation guideline provides a sequence of instruction for implementation in any manufacturing enterprises to build their design ontology. An implementation guideline to capture comprehensive design information, model design information for better reuse and share is unique for manufacturing enterprises & enterprise engineering literature. Without the guideline, implementation would be limited to those enterprises that could interpret the model and translate it into a sequence of events.

APPENDIX   A

PRODUCT AND PROCESS DESIGN
ONTOLOGY

Overview

This section describes the whole ontology of the products and processes design in terms of its components. As mentioned earlier that an ontology building effort includes identifying three basic components, they are as follows:

A1. Entities

A2. Relations

Detail descriptions of each component of the design ontology are listed in the following sections.

A.1 ENTITIES

0. ENTITY	
A building bloc in the design domain being modeled	
Attribute	**Description**
Name	The entity's name
Number	A unique qualifier for the activity
Parent	None
Documentation	User defined description of Entity
User-Attributes	User defined attributes for an Entity
Sub kinds	Design activity, Design object, and Relationship

1. DESIGN ACTIVITY	
Any process of transforming state of the environment that happens over time	
Attribute	**Description**
Name	The Activity 's name
Number	A unique qualifier for the activity
Parent	ACTIVITIES of which this Activity is a component (Entity)
Documentation	User defined description of the Activity
User-Attributes	User defined attributes for an Activity
Components	ACTIVITIES that make up this Activity. Synonymous with sub activities, children, etc.
Authorizer	The agent responsible for this ACTIVITY, usually The ACTOR is considered the process owner.
Objects	The OBJECTS (non resource) used by this Activity
Resources	The ACTOR that performs the Activity
Sub kinds	Component of kind

1.2 PRODUCT DESIGN
A process of transforming customer/design requirements into detail design specification using design resources and satisfying design constraints. This activity happens over time.

Attribute	Description
Name	Product design
Number	A unique qualifier for the activity
Parent	Design Activity
Documentation	User defined description of the Product design
User-Attributes	User defined attributes for Product design
Components	ACTIVITIES that make up this Activity. Such as design analysis, test and evaluation, validation and verification, etc.
Authorizer	The agent responsible for this ACTIVITY, usually the collaborative design team is considered the process owner and has authority over design process.
Objects	The objects (non resource) used by this Activity, Objects in this case are material and information.
Resources	The collaborative design team and machine that performs the Activity
Sub kinds	Formulation, Conceptual design, Embodiment design, and Detail design.

1.2.1 INDUSTRIAL DESIGN & FORMULATION	
A process of transforming customer/design requirements into design requirements using design resources and satisfying design constraints. This activity ensures feasibility of the design.	
Attribute	**Description**
Name	Industrial design & formulation
Number	A unique qualifier for the activity
Parent	Product Design
Documentation	User defined description of the Industrial design & formulation
User-Attributes	User defined attributes for Industrial design & formulation
Components	ACTIVITIES that make up this Activity. Such as identify requirements, set targets, identify evaluation criteria, identify constraints, etc.
Authorizer	The agent responsible for this ACTIVITY, usually the collaborative design team is considered the process owner and has authority over design process.
Objects	The objects (non resource) used by this Activity, Objects in this case are material and information.
Resources	The collaborative design team and machine that performs the Activity.
Sub kinds	Customer requirements analysis, reviews and feasibility analysis.

1.2.1.1 CUSTOMER REQUIREMENTS ANALYSIS	
A process of analyzing customer requirements to check the feasibility of the design and identify design requirements.	
Attribute	**Description**
Name	Customer Requirements Analysis
Number	A unique qualifier for the activity
Parent	Industrial Design & Formulation
Documentation	User defined description of Customer Requirements Analysis
User-Attributes	User defined attributes for Customer Requirements Analysis
Components	ACTIVITIES that make up this Activity. Such as identify requirements, set targets, identify evaluation criteria, identify constraints, etc.
Authorizer	The agent responsible for this ACTIVITY, usually the collaborative design team is considered the process owner and has authority over design process.
Objects	The objects (non resource) used by this Activity, Objects in this case are material and literature survey, market study, etc.
Resources	The collaborative design team and machine that performs the Activity.
Sub kinds	Component of kind

1.2.1.2 REVIEWS & FEASIBILITY ANALYSIS
A process of reviewing designs requirements to make sure that the design is feasible and manufacturable using design resources. This activity identifies engineering design needs.

Attribute	Description
Name	Reviews & Feasibility Analysis
Number	A unique qualifier for the activity
Parent	Industrial Design & Formulation
Documentation	User defined description of the Reviews & Feasibility Analysis
User-Attributes	User defined attributes for Reviews & Feasibility Analysis
Components	ACTIVITIES that make up this Activity. Such as identify functional requirements, identify company requirements, identify manufacturing requirements, etc.
Authorizer	The agent responsible for this ACTIVITY, usually the collaborative design team is considered the process owner and has authority over design process.
Objects	The objects (non resource) used by this Activity, Objects in this case are material and information.
Resources	The collaborative design team and machine that performs the Activity
Sub kinds	Component of kind

1.2.2 CONCEPTUAL DESIGN	
An abstract embodiment of a working principle, geometry, and material; a phase of design when the physical principles are selected.	
Attribute	**Description**
Name	Conceptual Design
Number	A unique qualifier for the activity
Parent	Product Design
Documentation	User defined description of the Conceptual Design
User-Attributes	User defined attributes for Conceptual Design
Components	ACTIVITIES that make up this Activity. Such as analysis of design needs, identify design tasks, concept generation, concept evaluation, concept selection, etc.
Authorizer	The agent responsible for this ACTIVITY, usually the collaborative design team is considered the process owner and has authority over design process.
Objects	The objects (non resource) used by this Activity, Objects in this case are material and information.
Resources	The collaborative design team and machine that performs the Activity
Sub kinds	Analysis of design needs, identify design tasks, concept generation, concept evaluation, and concept selection.

1.2.2.1 ANALYSIS OF DESIGN NEED	
A process of identifying and clarifying product's functional requirements in detail level.	
Attribute	**Description**
Name	Analysis Of Design Need
Number	A unique qualifier for the activity
Parent	Conceptual Design
Documentation	User defined description of the Analysis Of Design Need
User-Attributes	User defined attributes for Analysis Of Design Need
Components	ACTIVITIES that make up this Activity. Such as activity analysis, decomposition diagrams, functions structure, etc.
Authorizer	The agent responsible for this ACTIVITY, usually the collaborative design team is considered the process owner and has authority over design process.
Objects	The objects (non resource) used by this Activity, Objects in this case are material and information.
Resources	The collaborative design team and machine that performs the Activity
Sub kinds	Component of Kind

1.2.2.2 IDENTIFY DESIGN TASKS	
A process of identifying and organizing design tasks, which are necessary to conduct the conceptual deigns process.	
Attribute	**Description**
Name	Identify Design Tasks
Number	A unique qualifier for the activity
Parent	Conceptual Design
Documentation	User defined description of the Identify Design Tasks
User-Attributes	User defined attributes for Identify Design Tasks
Components	ACTIVITIES that make up this Activity. Such as identify design goal, build representative space, formulation of design knowledge background, etc.
Authorizer	The agent responsible for this ACTIVITY, usually the collaborative design team is considered the process owner and has authority over design process.
Objects	The objects (non resource) used by this Activity, Objects in this case are material and information.
Resources	The collaborative design team and machine that performs the Activity
Sub kinds	Component of Kind

1.2.2.3 CONCEPT GENERATION	
A process of creating alternatives or design concept, which is the earliest representation of a new product or of alternative approaches to designing a new product.	
Attribute	**Description**
Name	Concept Generation
Number	A unique qualifier for the activity
Parent	Conceptual Design
Documentation	User defined description of the Concept Generation
User-Attributes	User defined attributes for Concept Generation
Components	ACTIVITIES that make up this Activity. Such as use creative methods, use brainstorming, review existing products, create checklists, etc.
Authorizer	The agent responsible for this ACTIVITY, usually the collaborative design team is considered the process owner and has authority over design process.
Objects	The objects (non resource) used by this Activity, Objects in this case are material and information.
Resources	The collaborative design team and machine that performs the Activity
Sub kinds	Component of Kind

1.2.2.4 CONCEPT EVALUATION	
A process of evaluating all generated alternatives to find out the feasible and producible alternatives and predict their performances.	
Attribute	**Description**
Name	Concept Evaluation
Number	A unique qualifier for the activity
Parent	Conceptual Design
Documentation	User defined description of the Concept Evaluation
User-Attributes	User defined attributes for Concept Evaluation
Components	ACTIVITIES that make up this Activity. Such as conduct trade of analysis, conduct reviews, check feasibility and producibility, etc.
Authorizer	The agent responsible for this ACTIVITY, usually the collaborative design team is considered the process owner and has authority over design process.
Objects	The objects (non resource) used by this Activity, Objects in this case are material and information.
Resources	The collaborative design team and machine that performs the Activity
Sub kinds	Component of Kind

1.2.2.5 CONCEPT SELECTION	
A process of selecting best alternatives among available design alternatives.	
Attribute	**Description**
Name	Concept Selection
Number	A unique qualifier for the activity
Parent	Conceptual Design
Documentation	User defined description of the Concept Selection
User-Attributes	User defined attributes for Concept Selection
Components	ACTIVITIES that make up this Activity. Such as use Pugh's concept selection method or use weighted-rating method, etc.
Authorizer	The agent responsible for this ACTIVITY, usually the collaborative design team is considered the process owner and has authority over design process.
Objects	The objects (non resource) used by this Activity, Objects in this case are material and information.
Resources	The collaborative design team and machine that performs the Activity
Sub kinds	Component of Kind

1.2.3   EMBODIMENT DESIGN

A process of selecting detail specifications of product and it's part with the consideration of various aspect of design such as safety, manufacturability, assembly, environment, etc.

Attribute	Description
Name	Embodiment Design
Number	A unique qualifier for the activity
Parent	Product Design
Documentation	User defined description of the Embodiment Design
User-Attributes	User defined attributes for Embodiment Design
Components	ACTIVITIES that make up this Activity. Such as configure product, configure parts, configuration analysis & evaluation, select design variables, analyze alternatives, evaluate & refine configuration, etc.
Authorizer	The agent responsible for this ACTIVITY, usually the collaborative design team is considered the process owner and has authority over design process.
Objects	The objects (non resource) used by this Activity, Objects in this case are material and information.
Resources	The collaborative design team and machine that performs the Activity
Sub kinds	Configuration design and parametric design.

1.2.3.1 CONFIGURATION DESIGN
A process of selection and arrangement of features on a part; or the selection and arrangement of components on a product; a phase of design when geometric features are arranged and connected on a part, or standard components or types are selected for the architecture.

Attribute	Description
Name	Configuration Design
Number	A unique qualifier for the activity
Parent	Embodiment Design
Documentation	User defined description of the Configuration Design
User-Attributes	User defined attributes for Configuration Design
Components	ACTIVITIES that make up this Activity. Such as configure product, configure parts, configuration analysis & evaluation, etc.
Authorizer	The agent responsible for this ACTIVITY, usually the collaborative design team is considered the process owner and has authority over design process.
Objects	The objects (non resource) used by this Activity, Objects in this case are material and information.
Resources	The collaborative design team and machine that performs the Activity
Sub kinds	Configure product, configure parts, configuration analysis & evaluation.

1.2.3.1.1 CONFIGURE PRODUCT	
A process of determining the number & type of components, their specific functions, and their arrangements.	
Attribute	**Description**
Name	Configure Product
Number	A unique qualifier for the activity
Parent	Configuration Design
Documentation	User defined description of the Configure Product
User-Attributes	User defined attributes for Configure Product
Components	ACTIVITIES that make up this Activity. Such as determine number and types of components, determine specific functions, determine arrangements, etc.
Authorizer	The agent responsible for this ACTIVITY, usually the collaborative design team is considered the process owner and has authority over design process.
Objects	The objects (non resource) used by this Activity, Objects in this case are material and information.
Resources	The collaborative design team and machine that performs the Activity
Sub kinds	Component of Kind

1.2.3.1.2   CONFIGURE PARTS
A process of determining the number & type of geometric features, relative dimensions, and their arrangements.

Attribute	Description
Name	Configure Part
Number	A unique qualifier for the activity
Parent	Configuration Design
Documentation	User defined description of the Configure Part
User-Attributes	User defined attributes for Configure Part
Components	ACTIVITIES that make up this Activity. Such as determine number and types of geometric features, arrangements of parts, determine relative dimensions, selection of standard parts, etc.
Authorizer	The agent responsible for this ACTIVITY, usually the collaborative design team is considered the process owner and has authority over design process.
Objects	The objects (non resource) used by this Activity, Objects in this case are material and information.
Resources	The collaborative design team and machine that performs the Activity
Sub kinds	Component of Kind

1.2.3.1.3 ANALYZE & REFINE CONFIGURATION	
A process of determining configuration alternatives considering all aspect of design principles.	
Attribute	**Description**
Name	Analyze & Refine Configuration
Number	A unique qualifier for the activity
Parent	Configuration Design
Documentation	User defined description of the Analyze & Refine Configuration
User-Attributes	User defined attributes for Analyze & Refine Configuration
Components	ACTIVITIES that make up this Activity. Such as requirements analysis, design for assembly, design for manufacture, etc.
Authorizer	The agent responsible for this ACTIVITY, usually the collaborative design team is considered the process owner and has authority over design process.
Objects	The objects (non resource) used by this Activity, Objects in this case are material and information.
Resources	The collaborative design team and machine that performs the Activity
Sub kinds	Component of Kind

1.2.3.1.4   EVALUATE & SELECT BEST CONFIGURATION
A process of selecting best configuration after evaluating all available configuration alternatives.

Attribute	Description
Name	Evaluate & Select Best Configuration
Number	A unique qualifier for the activity
Parent	Configuration Design
Documentation	User defined description of the Evaluate & Select Best Configuration
User-Attributes	User defined attributes for Evaluate & Select Best Configuration
Components	ACTIVITIES that make up this Activity. Such as design reviews, use any weighted-rating method, etc.
Authorizer	The agent responsible for this ACTIVITY, usually the collaborative design team is considered the process owner and has authority over design process.
Objects	The objects (non resource) used by this Activity, Objects in this case are material and information.
Resources	The collaborative design team and machine that performs the Activity
Sub kinds	Component of Kind

1.1.3.2 PARAMETRIC DESIGN	
A phase of design that determines specific values for the design variables.	
Attribute	**Description**
Name	Parametric Design
Number	A unique qualifier for the activity
Parent	Embodiment Design
Documentation	User defined description of the Parametric Design
User-Attributes	User defined attributes for Parametric Design
Components	ACTIVITIES that make up this Activity. Such as select design variables, analyze alternatives, evaluate & refine configuration, etc.
Authorizer	The agent responsible for this ACTIVITY, usually the collaborative design team is considered the process owner and has authority over design process.
Objects	The objects (non resource) used by this Activity, Objects in this case are material and information.
Resources	The collaborative design team and machine that performs the Activity
Sub kinds	Select design variables, analyze alternatives, evaluate & refine configuration.

1.1.3.2.1 SELECT DESIGN VARIABLES	
A process of identifying design variables and determine values of those design variables.	
Attribute	**Description**
Name	Select Design Variables
Number	A unique qualifier for the activity
Parent	Parametric Design
Documentation	User defined description of the Select Design Variables
User-Attributes	User defined attributes for Select Design Variables
Components	ACTIVITIES that make up this Activity. Such as identify design variables, determine design constraints, select values for design variables, etc.
Authorizer	The agent responsible for this ACTIVITY, usually the collaborative design team is considered the process owner and has authority over design process.
Objects	The objects (non resource) used by this Activity, Objects in this case are material and information.
Resources	The collaborative design team and machine that performs the Activity
Sub kinds	Component of Kind

1.1.3.2.2 ANALYZE, REFINE & EVALUATE DESIGN VARIABLES	
A process of identifying feasible & optimal values design variables and determine their expected performance.	
Attribute	**Description**
Name	Analyze, Refine & Evaluate Design Variables
Number	A unique qualifier for the activity
Parent	Parametric Design
Documentation	User defined description of the Analyze, Refine & Evaluate Design Variables
User-Attributes	User defined attributes for Analyze, Refine & Evaluate Design Variables
Components	ACTIVITIES that make up this Activity. Such as predict product performance, check feasibility, check optimality, use weighted-rating method, etc.
Authorizer	The agent responsible for this ACTIVITY, usually the collaborative design team is considered the process owner and has authority over design process.
Objects	The objects (non resource) used by this Activity, Objects in this case are material and information.
Resources	The collaborative design team and machine that performs the Activity
Sub kinds	Component of Kind

1.2.4   DETAIL DESIGN

A phase of design that results in the preparation of a package of information that includes drawings and specifications sufficient to manufacture a product.

Attribute	Description
Name	Product design
Number	A unique qualifier for the activity
Parent	Design Activity
Documentation	User defined description of the Product design
User-Attributes	User defined attributes for Product design
Components	ACTIVITIES that make up this Activity. Such as design analysis, test and evaluation, validation and verification, etc.
Authorizer	The agent responsible for this ACTIVITY, usually the collaborative design team is considered the process owner and has authority over design process.
Objects	The objects (non resource) used by this Activity, Objects in this case are material and information.
Resources	The collaborative design team and machine that performs the Activity
Sub kinds	Formulation, Conceptual design, Embodiment design, and Detail design.

1.2.4.1 DESIGN ANALYSIS & TRADE OFFS	
A phase of design that predicts or simulates performance of each alternative, reiterating to assure that all the candidates feasible.	
Attribute	**Description**
Name	Design Analysis & Trade Offs
Number	A unique qualifier for the activity
Parent	Detail Design
Documentation	User defined description of the Design Analysis & Trade Offs
User-Attributes	User defined attributes for Design Analysis & Trade Offs
Components	ACTIVITIES that make up this Activity. Such as examine design parameters & their interactions, produce cost estimates, produce drawings, etc.
Authorizer	The agent responsible for this ACTIVITY, usually the collaborative design team is considered the process owner and has authority over design process.
Objects	The objects (non resource) used by this Activity, Objects in this case are material and information.
Resources	The collaborative design team and machine that performs the Activity
Sub kinds	Component of Kind

1.2.4.2 DETAIL FUNCTIONALITY ANALYSIS

A phase of design that determines the functional characteristics of each component using extensive analysis.

Attribute	Description
Name	Detail Functionality Analysis
Number	A unique qualifier for the activity
Parent	Detail Design
Documentation	User defined description of the Detail Functionality Analysis
User-Attributes	User defined attributes for Detail Functionality Analysis
Components	ACTIVITIES that make up this Activity. Such as stress analysis, finite element analysis, failure mode analysis, environmental stress analysis, etc.
Authorizer	The agent responsible for this ACTIVITY, usually the collaborative design team is considered the process owner and has authority over design process.
Objects	The objects (non resource) used by this Activity, Objects in this case are material and information.
Resources	The collaborative design team and machine that performs the Activity
Sub kinds	Component of Kind

1.2.4.3 DETAIL SYSTEMS DESIGN
A phase of design that determines the required systems specifications for the product to be produced.

Attribute	Description
Name	Detail Systems Design
Number	A unique qualifier for the activity
Parent	Detail Design
Documentation	User defined description of the Detail Systems Design
User-Attributes	User defined attributes for Detail Systems Design
Components	ACTIVITIES that make up this Activity. Such as supply chain and logistics design, packaging design, tools & fixture design, material handling design, etc.
Authorizer	The agent responsible for this ACTIVITY, usually the collaborative design team is considered the process owner and has authority over design process.
Objects	The objects (non resource) used by this Activity, Objects in this case are material and information.
Resources	The collaborative design team and machine that performs the Activity
Sub kinds	Component of Kind

1.2.4.4 FINAL DESIGN REVIEWS & VERIFICATION

A phase of design that identifies technical performance, risks, improvement of performance & process, and verifies design parameters.

Attribute	Description
Name	Final Design Reviews & Verification
Number	A unique qualifier for the activity
Parent	Detail Design
Documentation	User defined description of the Final Design Reviews & Verification
User-Attributes	User defined attributes for Final Design Reviews & Verification
Components	ACTIVITIES that make up this Activity. Such as reviews for technical performance, performance improvement, process & product improvement, etc.
Authorizer	The agent responsible for this ACTIVITY, usually the collaborative design team is considered the process owner and has authority over design process.
Objects	The objects (non resource) used by this Activity, Objects in this case are material and information.
Resources	The collaborative design team and machine that performs the Activity
Sub kinds	Component of Kind

1.2.4.5 TEST, EVALUATION & VALIDATION	
A process of evaluating each design specifications and validation through extensive testing and simulation.	
Attribute	**Description**
Name	Test, Evaluation & Validation
Number	A unique qualifier for the activity
Parent	Detail Design
Documentation	User defined description of the Test, Evaluation & Validation
User-Attributes	User defined attributes for Test, Evaluation & Validation
Components	ACTIVITIES that make up this Activity. Such as test, analyze & fix, environment test, accelerated life test, vendor test, user & field test, etc.
Authorizer	The agent responsible for this ACTIVITY, usually the collaborative design team is considered the process owner and has authority over design process.
Objects	The objects (non resource) used by this Activity, Objects in this case are material and information.
Resources	The collaborative design team and machine that performs the Activity
Sub kinds	Component of Kind

1.3 PROCESS DESIGN

Defining and designing a manufacturing process to accommodate the specific requirements of a given product while, meeting process quality and cost objectives.

Attribute	Description
Name	Process design
Number	A unique qualifier for the activity
Parent	Product & Process Design
Documentation	User defined description of the process design
User-Attributes	User defined attributes for process design
Components	ACTIVITIES that make up this Activity. Such as identifying processes, sequencing processes, test and evaluation, validation and verification, etc.
Authorizer	The agent responsible for this activity, usually the collaborative design team is considered the process owner and has authority over process design.
Objects	The objects (non resource) used by this Activity, Objects in this case are material and information.
Resources	The collaborative design team and machine that performs the Activity
Sub kinds	Production process planning, Verify and validate process

1.3.1 PRODUCTION PROCESS PLANNING	
A process of planning and determining process specifications to produce the designed products.	

Attribute	Description
Name	Production Process Planning
Number	A unique qualifier for the activity
Parent	Product and process design
Documentation	User defined description of the Production Process Planning
User-Attributes	User defined attributes for Production Process Planning
Components	ACTIVITIES that make up this Activity. Such as process capability analysis, process requirements analysis, facilities layout drawing, resource allocation, etc.
Authorizer	The agent responsible for this ACTIVITY, usually the collaborative design team is considered the process owner and has authority over process planning.
Objects	The objects (non resource) used by this Activity, Objects in this case are material and information.
Resources	The collaborative design team and machine that performs the Activity
Sub kinds	Component of kind

1.3.2 VERIFICATION AND VALIDATION OF PROCESSES	
A process of finalizing the production process specifications and release for production.	
Attribute	**Description**
Name	Verification and Validation of Processes
Number	A unique qualifier for the activity
Parent	Product and process design
Documentation	User defined description of the Verification and Validation of Processes
User-Attributes	User defined attributes for Verification and Validation of Processes
Components	ACTIVITIES that make up this Activity. Such as design analysis, test and evaluation, validation and verification, etc.
Category	Number designating whether the essence of this Activity is as a Category 1, 2, or 3 processes.
Authorizer	The agent responsible for this ACTIVITY, usually the collaborative design team is considered the process owner and has authority over Verification and Validation of Processes.
Authority	The AUTHORITY, which applies to this Activity
Objects	The objects (non resource) used by this Activity, Objects in this case are material and information.
Resources	The collaborative design team and machine that performs the Activity
Sub kinds	Component of kind

2. DESIGN OBJECT	
Any tangible or conceptual entity in the design domain	

Attribute	Description
Name	Design Object's Name
Number	A unique qualifier for the Design Object
Parent	Entity
Documentation	User defined description of the Design Object
User-Attributes	User defined attributes for a Design Object
Performs	The ACTIVITIES, which this DESIGN OBJECT is performing.
Sub kinds	Actor and Product

2.2 ACTOR	
A design expert, collaborative design team, or other entity capable of actively participating in an activity.	

Attribute	Description
Name	Actor's Name
Number	A unique qualifier for the Actor
Parent	Design Object
Documentation	User defined description of the Actor
User-Attributes	User defined attributes for an Actor
Performs	The ACTIVITIES, which this Actor is performing.
Sub kinds	Agent, Machine, and Design tool

2.2.1 DESIGN EXPERT

A human being entity that performs activities to achieve design goals. Its ability to actively participate in the planning process is what separates a DESIGN EXPERT from a MACHINE

Attribute	Description
Name	Design Expert 's name
Number	A unique qualifier for the Design Expert
Parent	Actor
Documentation	User defined description of the Design Expert
User-Attributes	User defined attributes for a Design Expert
Performs	The ACTIVITIES, which this DESIGN EXPERT is performing.
Sub kinds	Component of kind

2.2.2 MACHINE

A non-human entity, which has the capacity to carry out design functions.

Attribute	Description
Name	Machine 's name
Number	A unique qualifier for the Machine
Parent	Actor
Documentation	User defined description of the Machine
User-Attributes	User defined attributes for a Machine
Performs	The ACTIVITIES, which this MAHINE is performing.
Sub kinds	Component of kind

2.2.3   DESIGN TOOL

A non-human entity, which has the capacity to carry out design functions.

Attribute	Description
Name	Design Tool's name
Number	A unique qualifier for the Design Tool
Parent	Actor
Documentation	User defined description of the Design Tool
User-Attributes	User defined attributes for a Design Tool
Performs	The ACTIVITIES, which this DESIGN TOOL is performing.
Sub kinds	Component of kind

2.3 PRODUCT

Any passive object that can be used, created, or modified by an activity in the design domain. These are typically the inputs and outputs of a design process.

Attribute	Description
Name	Product 's name
Number	A unique qualifier for the Product
Parent	Design Object
Documentation	User defined description of the Product
User-Attributes	User defined attributes for a Product
Performs	The ACTIVITIES, which this PRODUCT is used to perform.
Sub kinds	Material, Information

2.3.1 MATERIAL

Any material object that can be used, created, or modified or consumed by an activity in the design domain (e.g. Raw materials, Subassemblies).

Attribute	Description
Name	Material 's name
Number	A unique qualifier for the Material
Parent	Product
Documentation	User defined description of the Material
User-Attributes	User defined attributes for a Material
Performs	The ACTIVITIES, which this MATERIAL is used to perform.
Sub kinds	Component of kind

2.3.2 INFORMATION

Any object that can be used, created, or modified by an activity and whose essence is to act as information in the design domain.

Attribute	Description
Name	Information 's name
Number	A unique qualifier for the Information
Parent	Product
Documentation	User defined description of the Information
User-Attributes	User defined attributes for a Information
Performs	The ACTIVITIES, which this INFORMATION is used to perform.
Sub kinds	Component of kind

A.2 RELATIONS

1. ASSIGNED TO		
Entity 1	**Entity 2**	**Definition**
Industrial design & formulation	Collaborative design team	Design activities like industrial design and formulation are assigned to collaborative design team.

2. BEGINS WITH		
Entity 1	**Entity 2**	**Definition**
Concept design	Concept generation	Defines the start point of design activities like concept design.

3. CONDUCTS		
Entity 1	**Entity 2**	**Definition**
Detail design analysis	Trade off studies	Describes the types of activities, tests, or analysis, which are needed to perform to do the products and processes design.

4. CONSISTS OF		
Entity 1	**Entity 2**	**Definition**
Industrial design & formulation	Formulation steps	Shows the components, ingredients, or part of design activities such as industrial design and formulation.

5. CONTROLS		
Entity 1	**Entity 2**	**Definition**
Design constraints	Design activities	Design constraints are entities, which limits the performance of design activities or control the way of design functions.

6. CHECKS FOR		
Entity 1	**Entity 2**	**Definition**
Design reviews & validation	Process improvement	As part of some design activities such as design reviews explores/ check for improvement in process & product or technical risks.

7. DEALS WITH		
Entity 1	**Entity 2**	**Definition**
Configuration design	Product architecture	Some design activities has to deal with some other concerns or tasks.

8. DEFINES		
Entity 1	**Entity 2**	**Definition**
Conceptual design	Design tasks	Conceptual design defines what are the design tasks and they will be performed.

9. DELIGATES TO		
Entity 1	**Entity 2**	**Definition**
Collaborative design team	Design expert	Collaborative design team delegates or assign some activities to individual design experts.

10. DESCRIBES		
Entity 1	**Entity 2**	**Definition**
Special part	Parts arrangement	A special part describes how it will interact with other parts and sequence of arrangements.

11. DETERMINES		
Entity 1	**Entity 2**	**Definition**
Parametric design	Part dimensions	Part dimension values are determined in the parametric design process.

12. DOCUMENTS		
Entity 1	**Entity 2**	**Definition**
Collaborative design team	Lessons learned	Collaborative design team has to document the lessons learned in the design process for future reuse.

13. ENDS AT		
Entity 1	**Entity 2**	**Definition**
Conceptual design	Concept selection	Defines the end point of design activities. Like conceptual design, it finishes its activities by completing the selection of alternatives.

14. EXAMINES		
Entity 1	**Entity 2**	**Definition**
Detail design	Parameter interaction	Detail design examines different aspects of parameter interactions.

15. HAS ACTIVITIES		
Entity 1	**Entity 2**	**Definition**
Detail design	Detail design activities	This relation designating the activities contained in the activity list. Detail design has list of activities, which are performed to the complete the activity.

16. HAS ASSIGNED		
Entity 1	**Entity 2**	**Definition**
Design resources	Planning design tasks	Defines the planning design assigned to an Actor. Most of the design resources can be assigned to particular design activities.

17. HAS CHILDREN		
Entity 1	**Entity 2**	**Definition**
Product & process design	Process design	The relation in which one design activity is a component of another.

18. HAS COMPONENTS		
Entity 1	**Entity 2**	**Definition**
Technical requirements	Components of technical requirements	A requirement is a part of another requirement.

19. HAS CONTROL OVER		
Entity 1	**Entity 2**	**Definition**
Design constraints	Process design	Design constraints have control over the design process.

20. HAS INPUT		
Entity 1	**Entity 2**	**Definition**
Industrial design & formulation	Formulation inputs	Some design activities has inputs, whch is transformed into outputs due to the result of the activity.

21. HAS MECHANISM		
Entity 1	**Entity 2**	**Definition**
Detail design	Design resources	This relation describes the means of performing the design activities. Design resources provide the mechanism to perform detail design.

22. HAS PARENT		
Entity 1	**Entity 2**	**Definition**
Industrial design & formulation	Product design	Industrial design and formulation has parent activity. It is a decomposition of product design.

23. HAS PERFORMANCE		
Entity 1	**Entity 2**	**Definition**
Actor	Collaborative design team performance	Define the performance (either as capability or actual performance), which an actor possesses.

24. HAS OUTPUT		
Entity 1	**Entity 2**	**Definition**
Conceptual Design	Conceptual design outputs	Conceptual design produces output like conceptual design outputs.

25. INCLUDES		
Entity 1	**Entity 2**	**Definition**
Formulation steps	Reviews & feasibility analysis	Reviews & feasibility analysis contained in the formulation steps.

26. INCLUDES ACTIVITIES		
Entity 1	**Entity 2**	**Definition**
Process design	Process design activities	Any parent activity includes list of children activities or sub activities.

27. IS ASSIGNED		
Entity 1	**Entity 2**	**Definition**
Design resources	Design activities	Design resources has been assigned to various design activities.

28. IS PART OF		
Entity 1	**Entity 2**	**Definition**
Design for functionality	Design for close fit	Design for close fit is part of design for functionality.

29. MEETS		
Entity 1	**Entity 2**	**Definition**
Concept evaluation & selection	Design requirements	Concept evaluation & selection must satisfy all design requirements to have a feasible design.

30. PERFORMS		
Entity 1	**Entity 2**	**Definition**
Concept design	Design need analysis	Design activities or Actor perform some task to complete the activities.

31. PRODUCES		
Entity 1	**Entity 2**	**Definition**
Detail design	Design specifications	Detail design produces design specifications for production processes.

32. TRANSFORMS		
Entity 1	**Entity 2**	**Definition**
Concept design	Refined best concepts	Concept design transforms customer requirements into refined best concepts/ alternatives.

33. USES		
Entity 1	**Entity 2**	**Definition**
Configuration design	Design resources	Define the objects/ design resources used by an activity of Actor.

34. USES AS INPUT		
Entity 1	**Entity 2**	**Definition**
Industrial design & formulation	Customer requirements	This relation defines the objects/inputs used by an activity.

35. USES AS OUTPUT		
Entity 1	**Entity 2**	**Definition**
Detail design	Drawings	This relation defines the objects/outputs used by an activity.

APPENDIX   B

QUESTIONNAIRES & IMPLEMENTATION
GUIDELINE

Overview

This section describes the application of ontology development methodology, which was developed in chapter four in any manufacturing enterprises. To make the application easier an implementation guideline was developed and presented in this section. This implementation guideline consists of instruction with brief descriptions of each step, so that any design related organization can implement the methodology without browsing the whole book. A questionnaire and interview kit were developed to collect design information for the demonstration purpose and presented in the following section. This section describes the following items in detail.

B1. Interview Kit and Questionnaire

B2. Implementation Guidelines

B.1 INTERVIEW KIT & QUESTIONNAIRE

"INTERVIEW KIT" For Building Product and Process Design Ontology

Interview Kit is document for conducting formal interviews with the participation companies using questionnaires or other supporting materials. A standard interview kit contains the following.

- Introduction
- Research Overview
- Problem Statement
- Research Objective
- Purpose of Interview
- Interview Methodology
- Selection of Interviewees
- Topics of Interview
- Types of Information Sought
- Data Collection Method
- Detail Questionnaire
- Follow-up

Introduction

Introduce yourself and your data collection projects. Mention your affiliation and relation with your referees. Also talk about your project such as you are developing a methodology for documenting product and process design knowledge.

Research Overview

Provide an overview of the research you are conducting. In this particular case a brief overview is as follows. The technical term "ontology of product and process design", is a formal description of what is known about product and process design. The result of this work will be a method to describe what is known about product and process design in a particular setting like a single company.

Problem Statement

State your problems in detail. This gives an idea to the company you are conducting research about your problems. For this effort, problem statements are-

- To reduce the cost and product development time, manufacturing enterprises need to share design information and use pre developed proven design templates.

- Many researchers contributed towards the institutionalization of design process for manufacturing enterprises but still there is no comprehensive design library for manufacturing enterprises.

- Ontologies are becoming increasingly important because they provide the critical semantic foundation for many rapidly expanding fields of knowledge. They are very useful for knowledge reuse, knowledge sharing, and enterprise modeling.

- Manufacturing enterprise needs a structured methodology to model their design knowledge for future reuse and sharing among other enterprises and/or within various departments.

Research Objectives

State your objectives. Mention what you are trying to achieve and how that will help the participating companies or to the research community. In this case, the objective is to develop a methodology for creating an ontology of product and process design for manufacturing enterprises, which enable them to be competitive in the present marketplace by reducing design lead-time and cost of design. A cross consistency matrix will be developed to measure the level of consistency and accuracy of information captured and represented by the ontology.

This methodology will help manufacturing enterprises to

- Capture the knowledge of their interest

- Share common understanding of information across the orgs

- Enable reuse of captured knowledge & domain knowledge

- Make explicit domain assumptions

- Separate domain knowledge from the operational knowledge

- Analyze domain knowledge

Purpose of Interview

Describe the reason for conducting this interview. You can tell that you are in the process of developing the methodology and expect to finish it very soon. After you are finished with development, you need to validate the methodology using several real world cases. The Information you will be collecting through this interview will be used to validate your methodology and will not be shared with any body else.

Interview Methodology

Before collecting any information, it is necessary to mention about the procedure of your interview methodology. This gives an idea to the interviewee and facilitates better communication. Also mention about the time and resources you need from the participating companies and what kind of & how detail information you are looking for. In this effort, interview methodology mentions about asking access to design team members, mid level managers, shop floor level production managers, and workers. This initial plan is to interview approximately 10 people from the participating company. The interview process is a two sessions process and does not long more than two hours. Which means, the whole interview process will not more than forty hours of time.

Mention that you are particularly looking for design information of their product and process design. This information includes design process in terms of activities and

230

its sequences, their interpretation of standard design terms, source of their design information, the way they do the design, and so on (NOT what tools and techniques they use). You also tell them that you will collect the above-mentioned information through interviews, questionnaires, and personal observations.

DETAIL QUESTIONNAIRE

The result of this interview is used only for educational purpose and to improve the body of product and process design knowledge. Individual information of this interview will not be revealed to the management.

1. EMPLOYEE INFORMATION

Your designation in the job:	
Age:	
How long are you working for this company?	
How long are you working for this particular job?	
How many product you have designed in this job?	
How many process designs you were involved with?	
Have you heard about design Ontology?	
What minimum formal training and/or educational level is required to do your job?	

2. JOB INFORMATION

Position Summary: *(State the main purpose of your job)*

In a few words, please provide a brief description of your job, focusing on the overall purpose and key objectives required for your job.

Responsibilities/Activities:

List all job responsibilities in order of importance. For each duty, indicate the percentage of time devoted to each duty.

Duties	Percentage of Time (%)
1.	
2.	
3.	
4.	
5.	
6.	

3. BUILDING DESIGN ONTOLOGY

Design Ontology is a formal description of what is known about product and process design. It is hierarchically structured set of terms for describing a design domain that can be used as a skeletal foundation for a knowledge base.

Q 1. Who is responsible for your Product and process design? Or what consists of your collaborative design team? (Title of position and department only)

Q 2. Please select your overall design process among the following standard design
processes.

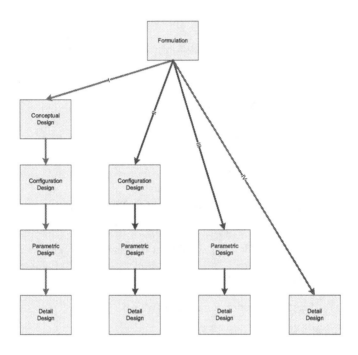

- Original Product Design (If you have checked, please go to question # 3)

- Part Configuration Design (If you have checked, please go to question # 9)

- Variant design (If you have checked, please go to question # 9)

- Selection Design (If you have checked, please go to question # 14)

Q 3. What do you consider your major steps in your design process?

Q 4. What do you do in your conceptual design process? Or what are the activities in your conceptual design process?

Q 5. Please define the important terms you just have mentioned in question # 4? These definitions are based on your experience in the job.

Q 6. What is the sequential relationship (if any) among theses conceptual design activities (Mentioned in question # 4)?

Q 7. How do you rank these activities identified in Q # 4, according to the importance of this function in the design process? (9 for most important and 0 for not important at all)

List of activities	Rank (0 – 9)
I.	
II.	
III.	
IV.	
V.	
VI.	
VII.	
VIII.	
IX.	
X.	
XI.	
XII.	
XIII.	
XIV.	
XV.	
XVI.	
XVII.	
XVIII.	
XIX.	
XX.	
XXI.	
XXII.	
XXIII.	
XXIV.	
XXV.	

Q 8. How are these activities related to each other?

Please use the following symbols;

    xx for strong relationship

    x   for moderate relationship and

    o   for no relationship

List of Conceptual	Activity # 1	Activity # 2	Activity # 3	Activity # 4	Activity # 5	Activity # 6	Activity # 7	Activity # 8	Activity # 9	Activity # 10	Activity # 11	Activity # 12	Activity # 13	Activity # 14	Activity # 15	Activity # 16	Activity # 17	Activity # 18	Activity # 19	Activity # 20
Design Activities																				
Activity # 1																				
Activity # 2																				
Activity # 3																				
Activity # 4																				
Activity # 5																				
Activity # 6																				
Activity # 7																				
Activity # 8																				
Activity # 9																				
Activity # 10																				
Activity # 11																				
Activity # 12																				
Activity # 13																				
Activity # 14																				
Activity # 15																				
Activity # 16																				
Activity # 17																				
Activity # 18																				
Activity # 19																				
Activity # 20																				

Q 9. What do you do in your Embodiment/Preliminary design process? Or what are the activities in your Embodiment/Preliminary design process?

Q 10. Please define the important terms you just have mentioned in question # 9? These definitions are based on your experience in the job.

Q 11. What is the sequential relationship (if any) among theses Embodiment/ Preliminary design activities (Mentioned in question # 9)?

Q 12. How do you rank these activities identified in Q # 9, according to the importance of this function in the design process? (9 for most important and 0 for not important at all)

List of activities	Rank (0 – 9)
I.	
II.	
III.	
IV.	
V.	
VI.	
VII.	
VIII.	
IX.	
X.	
XI.	
XII.	
XIII.	
XIV.	
XV.	
XVI.	
XVII.	
XVIII.	
XIX.	
XX.	
XXI.	
XXII.	
XXIII.	
XXIV.	
XXV.	

Q 13. How are these activities related to each other?

Please use the following symbols;

    xx for strong relationship

    x   for moderate relationship and

    o   for no relationship

List of Embodiment Design Activities	Activity # 1	Activity # 2	Activity # 3	Activity # 4	Activity # 5	Activity # 6	Activity # 7	Activity # 8	Activity # 9	Activity # 10	Activity # 11	Activity # 12	Activity # 13	Activity # 14	Activity # 15	Activity # 16	Activity # 17	Activity # 18	Activity # 19	Activity # 20
Activity # 1																				
Activity # 2																				
Activity # 3																				
Activity # 4																				
Activity # 5																				
Activity # 6																				
Activity # 7																				
Activity # 8																				
Activity # 9																				
Activity # 10																				
Activity # 11																				
Activity # 12																				
Activity # 13																				
Activity # 14																				
Activity # 15																				
Activity # 16																				
Activity # 17																				
Activity # 18																				
Activity # 19																				
Activity # 20																				

Q 14. What do you do in your detail design process? Or what are the activities in your detail design process?

Q 15. Please define the important terms you just have mentioned in question # 14? These definitions are based on your experience in the job.

Q 16 What is the sequential relationship (if any) among these detail design activities (Mentioned in question # 14)?

Q 17. How do you rank these activities identified in Q # 14, according to the importance of this function in the design process? (9 for most important and 0 for not important at all)

List of activities	Rank (0 – 9)
I.	
II.	
III.	
IV.	
V.	
VI.	
VII.	
VIII.	
IX.	
X.	
XI.	
XII.	
XIII.	
XIV.	
XV.	
XVI.	
XVII.	
XVIII.	
XIX.	
XX.	
XXI.	
XXII.	
XXIII.	
XXIV.	
XXV.	

Q 18. How are these activities related to each other?

Please use the following symbols;

    xx for strong relationship

    x   for moderate relationship and

    o   for no relationship

List of Detail Design Activities	Activity # 1	Activity # 2	Activity # 3	Activity # 4	Activity # 5	Activity # 6	Activity # 7	Activity # 8	Activity # 9	Activity # 10	Activity # 11	Activity # 12	Activity # 13	Activity # 14	Activity # 15	Activity # 16	Activity # 17	Activity # 18	Activity # 19	Activity # 20
Activity # 1																				
Activity # 2																				
Activity # 3																				
Activity # 4																				
Activity # 5																				
Activity # 6																				
Activity # 7																				
Activity # 8																				
Activity # 9																				
Activity # 10																				
Activity # 11																				
Activity # 12																				
Activity # 13																				
Activity # 14																				
Activity # 15																				
Activity # 16																				
Activity # 17																				
Activity # 18																				
Activity # 19																				
Activity # 20																				

B.2 IMPLEMENTATION GUIDELINES

Overview

This implementation guideline consists of instruction with brief descriptions of each step, so that any design related organization can implement the methodology without browsing the whole book. Most of the steps are sequential and well organized. In the following sections these steps are described elaborately.

Structure of the Methodology

In developing design ontology, a well-structured methodology is essential to capture appropriate design knowledge. This research methodology is constructed to meet the particular design needs. The steps of the methodology are slightly different from other ontology building approach to tailor the capture of design knowledge. It has nine major steps. They are as follows.

1. Determine the domain & Scope of the ontology

2. Check availability of existing ontologies

3. Organize the project

4. Collect and Analyze Data

5. Develop Initial Ontology

6. Refine and Validate Ontology

7. Check consistency & accuracy of ontology

8. Collect additional data and analyze data

9. Incorporate lessons learned and publish ontology

The structure of this design methodology for ontology development process is shown in following figure. Two fundamental differences between this particular design ontology and an ordinary ontology are as follows. This methodology is capable to re-use already developed design ontology as is or with minor change. Another significance of this methodology is to check the consistency and accuracy of captured information, which is vital for design ontology. It also publishes the developed design ontology into shared ontology repository for others to re-use. As mentioned earlier, design process is almost similar for different products but very important for each products and processes. A common repository of design ontologies is very useful for many manufacturing enterprises for their design needs.

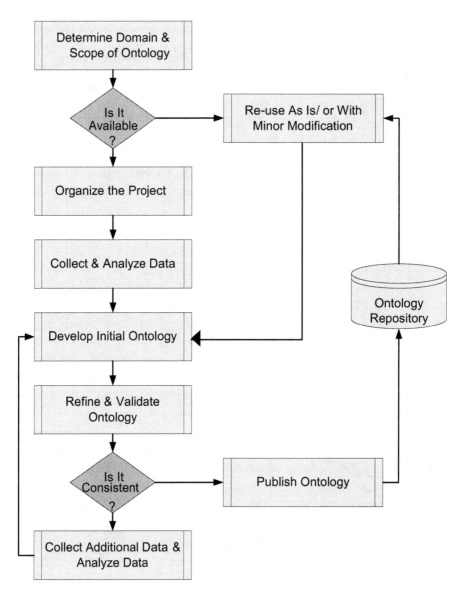

Figure B.1. Structure of DKAP Methodology

Implementation Steps of the Methodology

The implementation steps of the methodology are discussed in the following sections along with an example case.

Step 1. Determine the domain & Scope the ontology: This activity will establish the purpose, viewpoint, and context for the ontology development project and assign roles to the team members. The purpose statement provides a "completion criteria" for the ontology description capture effort. The purpose is usually established by a list of 1) statements of objectives for the effort, 2) statements of needs that the description must satisfy, and 3) questions or findings that need to be answered. For example, the purpose statement for this effort is: *"To develop design ontology of a generic product and process design".*

Once the purpose of the effort has been characterized, it is possible to define the context of the project in terms of 1) the scope of coverage, and 2) the level of detail for the ontology development effort. The scope defines the boundaries of the description development effort, and specifies which parts of the systems need to be included and which are to be excluded.

Establishing viewpoints is important to develop the ontology. It is related to the purpose of development. For instance, collaborative design team will normally use design ontology, hence it is appropriate to establish viewpoints with respect to collaborative design team. Nevertheless, the role of differing viewpoints on the outcome of ontology capture efforts is an important one. The differences in viewpoints are often reflected in different aspects of the ontology such as the specification of the level of detail of the description capture. Table B.2.1 shows an IDEF 5 form of ontology description summary including purpose, context and viewpoints.

Table B.1. Definition of the Ontology Development Project

Ontology Description Summary Form			
Project: Automobile Design Ontology	**Analyst:** Md Sarder	**Reviewer:** Don Liles	**Document** **Number:**
Version: 1	**Date:** 9/12/2005	**Date:** X/X/2005	
Purpose: To develop an ontology of the Product & Process Design domain for Automobile manufacturing enterprises. The resulting description must serve 1) as a knowledge repository for Company A's design system integration project and 2) as a reference model for Automobile industry as a whole.			
Context: The information acquired must be sufficient to organize design activities, specify precedence relationships, and supports world-class design procedures.			

Step 2. Check availability of existing ontologies: It is almost always worth considering what someone else has done and checking refinement and extends existing

sources for design domain and task. There is no valid reason to expend resources to build an ontology, which is already available. In some cases, a similar kind of ontology can be derived from the available one. Reusing existing ontologies may be a requirement if the system needs to interact with other applications that have already committed to particular ontologies or controlled vocabularies (Natalya & Mc Guinness, 2001). Many ontologies are already available in electronic form and can be imported into an ontology-development environment that someone is using. The formalism in which an ontology is expressed often does not matter, since many knowledge-representation systems can import and export ontologies. Even if a knowledge-representation system cannot work directly with a particular formalism, the task of translating an ontology from one formalism to another is usually not a difficult one.

Step 3. Organize the project: This activity will set different task to be performed to build the new ontology after checking that there are no ontology available to reuse as is or with minor change. Some of the tasks are to form Development Team, break down the tasks, assign team members to specific tasks, etc.

An important initial step in developing an ontology description is the formation of a development team. Each member of the team will perform a particular role in the development effort. Individuals who are involved in the modeling may each fulfill several roles, but each role is dealt with distinctly and should be clearly separated in the minds of the participants. The following are the sample roles assumed by the ontology development project personnel:

i.   Project Leader: This administrative role is responsible for overseeing and guiding the entire ontology development effort.

ii. Analyst/Knowledge Engineer: Personnel with ontology development expertise who will be the primary developers of the ontology description fill this technical role.

iii. Domain Expert: This role characterizes the primary sources of knowledge from the application domain of interest. Persons filling this role will provide insights about the characteristics of the application domain that are needed for extracting the underlying ontological knowledge.

iv. Team Members: All persons involved with the ontology description project.

A Work Breakdown Structure (WBS) is a results-oriented family tree that captures all the work of a project in an organized way. It is often portrayed graphically as a hierarchical tree; however, it can also be a tabular list of "element" categories and tasks or the indented task list that appears in the Gantt chart schedule. Following figure shows the WBS of building design ontology.

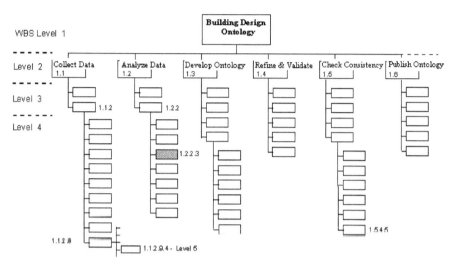

WBS Level 1

Level 2

Level 3

Level 4

Figure B.2. WBS Structure of Building Design Ontology

The WBS should be designed with consideration for its eventual uses. WBS design should try to achieve certain goals:

- Be compatible with how the work will be done and how schedules will be managed

- Give visibility to important or risky work efforts

- Allow mapping of requirements, plans, testing, and deliverables

- Foster clear ownership by project leaders and team members and

- Provide data for performance measurement and historical databases

Once a complete WBS is constructed, team members are assigned against each individual task to ensure the progress of the ontology building effort.

Step 4. Collect and Analyze Data: This activity will acquire the raw data needed for ontology development and analyze the data to facilitate ontology extraction. The definition of viewpoint, context, and purpose sets the stage for the data-gathering phase of the ontology captures effort. One of the problems in data collection is determining the appropriate sources of data. Various research experiences indicate that the main data sources are the domain expert and documents relevant to the circumscribed ontology. Regardless of the data collection methods used, it is important at this stage to establish an action plan for collecting data pertinent to the purpose and viewpoint of the model. Once collected, each piece of collected data must be traceable to its source. Traceability of source material is important because it is the data, which provides objective evidence for the basic ontological structures that are later isolated from this data.

Three important support documents can be used to facilitate source data traceability:

  i. Source Material Log: A document that serves as the primary index to all source material used in the project. This log lists all the materials used in the project. This list contains the name of the source materials, sources, which collected and when collected and shown in the Table B.2.2. Each entry of this document is then described using separate forms called Source Material Description Form shown in Table B.2.3.

Table B.2. Source Material Log for Building Design Ontology

Source Material Log				
Project: Designing Product/process ontology			Analysts:	
Source material #	Source material name	Collecte d from	Collecte d by	Date of Collecti on
SM # 1	"Operate a Small Integrated Manufacturing Enterprise" by Don Liles, ARRI, 1998	---------	Sarder	12/15/04
SM # 2	"Product Development and Design for Manufacturing" by John W. priest & Jose Sanchez, Marcel Dekker, Inc. New York, 2001	---------	Sarder	04/21/04
SM # 3	"Collaborative Evaluation of Early Design Decisions and Product Manufacturability" by S. D. Kleban et el, Proceedings of the 34th Hawaii International Conference on System Sciences - 2001	---------	Sarder	6/23/04
SM # 4	"Complexity and learning behaviors in product innovation" by Ross Chapman and, Paul Hyland. Technovation 24, 2004	---------	Sarder	8/24/04
SM # 5	"Coordination at different stages of product design process" by Antonio J Bailetti et el, R&D Management 28, 4, 1998	---------	Sarder	12/14/04
SM # n	---	--------	--------	--------

Table B.3. Source Material Description Form for Building Design Ontology

Source Material Description Form
Project: Product & Process Design Ontology
Analysts: MD Sarder
Source material #: SM # 3
Source material name: Collaborative Evaluation of Early Design Decisions and Product Manufacturability" by S. D. Kleban et el, Proceedings of the 34th Hawaii International Conference on System Sciences – 2001
Purpose: To record the relevant source statements that help individuate ontology elements in the product & process design domain.
Comments: This source material concerns early design stage and manufacturing of goods.
Abstract: In manufacturing, the conceptual design and detailed design stages are typically regarded as sequential and distinct. Decisions made in conceptual design are often made with little information as to how they would affect detailed design or manufacturing process specification. Many possibilities and unknowns exist in conceptual design where ideas about product shape and functionality are changing rapidly. Few if any tools exist to aid in this difficult, amorphous stage in contrast to the many CAD and analysis tools for detailed design where much more is known about the final product. The Materials Process Design Environment (MPDE) is a collaborative problem solving environment (CPSE) that was developed so geographically dispersed designers in both the conceptual and detailed stage can work together and understand the impacts of their design decisions on functionality, cost and manufacturability.
Terms Supported: T#2, T#5, T#12, T#15
Statements supported: SS#3, SS#5

ii.   Statement Pool: This document records meaningful statements made by different individuals, as well as statements extracted from source documents during the ontology development effort. An example statement about the engineering design is: "Engineering Design activities result in recommended manufacturing specifications that satisfy the customer's functional performance requirements and manufacturing constraints". Table B.2.4 shows an example of a source statement pool. Table B.2.5 shows source statement description form, which describes each statement in detail.

Table B.4. Source Statement Pool for Building Design Ontology

Source Statement Pool		
Project: Product & Process Design Ontology		Analysts: MD Sarder
Source Statement #	Source Statement	Supported by
SS # 1	Engineering Design activities result in recommended manufacturing specifications that satisfy the customer's functional performance requirements and manufacturing constraints.	MD Sarder
SS # 2	Resources may be classified as personnel, computer systems, and facilities.	MD Sarder
SS # 3	---	--------------------
SS # n	---	--------------------

Table B.5. Source Statement Description Form for Building Design Ontology

Source Statement Description Form		
Project: Product & Process Design Ontology	Analysts: MD Sarder	
Source Statement #: SS#1	Statement #S Evolved To:	Status: *Active / Retired* *Original / Derived*
Source Material #: SM#2	Statement #S Derived From:	
Source Statement: Engineering Design activities result in recommended manufacturing specifications that satisfy the customer's functional performance requirements and manufacturing constraints.		Supported by:
Version 1: Engineering design deals with the target specifications that will meet customer's functional requirements.		Supported by:
Version 2:		Supported by:
Version 3:		Supported by:
Comments:		

    iii.   Term Pool: The Term Pool alphabetically records all the meaningful terms relevant to the ontology building effort. Terms would typically connote kinds/instances of kinds, and relations/instances of relations. In this example, a term pool would include such terms as engineering design, embodiment design, rapid prototyping, etc. Table B.2.6 shows a sample term pool and Table B.2.7 shows the description of each term.

Table B.6. Term Pool for Building Design Ontology

Term Pool				
Project: Product & Process Design Ontology			Analysts:	
Term #	Term	Source Statement Reference	Source Material Reference	Support list
Term # 1	Engineering Design	SS#1	SM#2	MS
Term # 2	Embodiment Design	SS#2	SM#13	DS, MS
Term # 3	------------------	--------------------	------------------------	------------
Term # n	------------------	--------------------	------------------------	------------

Table B.7. Term Description for Building Design Ontology

Term Description Form		
Project: Product & Process Design Ontology		Analysts:
Term #	Term	Description
Term#1	E. Design	Engineering design is process of translating customer requirement into design specifications.
Term#2	Resource	Resources are objects/personnel that are consumed, used, or required to perform activities and tasks. Resources play an enabling role in processes.
Term # 3	----------------	--
Term # n	----------------	--

The objective of data analysis of the ontology development process is to analyze source material that has been collected and construct an initial characterization of the ontology. This task is performed by the knowledge engineer/analyst closely teaming with the domain expert. This task will typically involve the activities such as identify relevant design processes, list the objects of interest in the domain, examine boundary objects for boundary refinement, etc.

     i.   Identify relevant design activities

In a product and process environment, the natures of many kinds of things in a domain, especially the important relations they bear to other entities in the enterprise, are revealed not so much by examination of those entities but the roles they play situated in the processes in which they figure. Hence, the first objective in ontology building in product and process design environment is to capture the relevant design activities in which the ontological elements of the domain participate. For example, in a design situation, these will include conceptual design, embodiment design, detail design, design for manufacturing, design for safety, life testing, prototyping, etc. These then provide the necessary contextual information for the construction of accurate and complete domain ontologies.

ii.   List the objects of interest in the domain

Several objects will be fairly obvious from an initial study of the activity descriptions resulting from the previous step. Other objects will be identifiable from the source data such as the Statement Pool and the Term Pool. For example, the different kinds of design aspects, decision analysis, methods, tools, and fixtures that are associated with product and process design will be obvious ontology candidates for design ontology. The viewpoint and context statements constructed earlier in the development process will guide the level of detail that needs to be employed to develop this list.

iii.   Examine boundary objects for boundary refinement

The initial boundaries defined in the context statement may need to be redrawn to facilitate better conceptual structuring of the ontology.

Boundaries are often expanded to accommodate important objects that were earlier on the boundary. For example, consider a Plotter machine that is used in the design processes for ABC's design. Suppose that "drawing equipment" were initially excluded from the scope of the project. Suppose further that there are eleven other kinds of drawing machines that are used to draw and print design drawings at ABC. At this point, the boundaries are redrawn to explicitly include drawing equipment as part of the design ontology.

iv.   Partition the domain into subsystems

Systems are defined as collections of physical and/or conceptual objects that work together for a common purpose. Organizing ontologies by the systems provides a clear conceptual framework for subsequent analysis of ontological knowledge. It is therefore important to partition the focus domain into clearly delineable subsystems early in the ontology development process. This design ontology building method provides a graphical language that supports the conceptual activities such as representation of a system at varying levels of abstraction.

Step 5. Develop Initial Ontology: This activity is to develop a preliminary ontology from the acquired data. In the previous step relevant data was collected and analyzed them to use in this step. In this step there are a series of task according to tool to be selected. For IDEF 5 tool, the task could be identifying Proto Properties, proto Relations, and proto Kinds, classify Kinds, Properties, and Relations, etc.

260

i.   *Identify Proto Properties and Proto Kinds*

Properties are the characteristics that hold of objects in the real world. Examples of properties are weight, color, age, shape, etc. In this ontology building method, concepts are initially catalogued as "proto" concepts, that is, they are tentative and subject to further inquiry before final change of status by eliminating the "proto" prefix. Thus, potential candidates for "properties" in the ontology are initially called "proto properties." Similarly, there are "proto kinds" and "proto relations" (described later in this section). Proto property identification usually occurs concurrently with proto kind identification. This is because kinds are usually individuated on the basis of the properties that they exhibit. Listing properties is a relatively straightforward task because properties are readily observable and are often measurable.

A *proto kind* is the result of a preliminary attempt at individuating a kind. This task essentially consists of associating the objects identified in the data analysis activity with the proto properties identified. It may be instructive to perform this association process in two stages. First, the association is carried up to the point where the proto kind can be clearly distinguished from any other proto kind, that is, the proto kinds have a basis for being uniquely individuated. Properties that contribute to the uniqueness of a kind are candidate-defining properties. Defining properties stipulate necessary conditions for membership to a kind. Once the defining properties are identified, the remaining properties (non-defining) that are used to characterize the kinds in greater detail are associated with the kinds. At this stage of the analysis, it often becomes clear which proto kinds represent genuine kinds, where two terms have been used to indicate the same kind, where the same term is used to indicate distinct kinds, and so on. Once the characterization of a proto kind is relatively complete, it is converted to a kind. That is, classification as a "proto" concept is no longer necessary in view of the evidence that supports the concept.

ii.   Classify Kinds and Kind Hierarchy

Deciding whether a particular concept is a Kind in an ontology or an individual instance depends on what the potential applications of the ontology are. In case of design ontology, recognizing the multiplicity of classification mechanisms in different domain areas, this method provides a range of different classification relations to aid domain experts to identify kinds & sub kinds. Depending on the context of use, the subkind (classification) relation can be categorized under three headings as described in the following.

a. Generalization-specialization: The generalization classification *Is-a* relation links a general kind with a specialization of the kind. For example, a *conceptual design* kind is a specialization of a *design steps* kind. This type of classification relation is widely used in a variety of different application domains.

b. Natural kind classification: Ontologies of physical objects are classified using the "kind classification" meaning of the Is-a relation. This relation, often dubbed "A Kind of Object" (AKO), bestows the distinguished status of kind hood to the related objects. Often there are no necessary and sufficient conditions for entry into a particular kind, and objects just "are" (i.e., by definition), or happen to be, of particular kinds. The AKO relation is used predominantly for classifying natural objects and natural phenomena. For example, a design software is a kind of design resources.

c. Description classification: Description classification relations are used to define one object kind in terms of another. This type of Is-a relation is particularly useful for describing abstract object kinds. For example, the assertion that "A square is a rectangle" is a concise way of asserting that "a square is a rectangle with four equal sides." Here, the description of the rectangle is used to define the concept of the square, that is, the square description subsumes the description of the rectangle.

Following figure illustrates the classification of design resources. Typically, design resources can be categorized as collaborative design team, equipment, and tools. Collaborative design team consists of design experts from various disciplines. Equipment can be further categorized as design testing machine, rapid prototype machine, and material storage device. A tool can be design software and decision analysis tool.

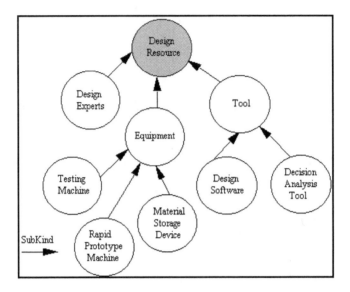

Figure B.3. Design Resource Entity

Once kinds are identified, they need to be placed in a hierarchy of order. There are several possible approaches in developing a kind hierarchy (Natalya and Mc Guinness, 2001):

- A top-down development process starts with the definition of the most general concepts in the domain and subsequent specialization of the concepts. For example, one can start with creating kinds for the general concepts of product and process design. Then he/she specializes the design Kind by creating some of its sub kinds: conceptual design, embodiment design, and detail design. One can further categorize the detail design, for example, into part analysis, field-testing, FEMA analysis, and so on.

- A bottom-up development process starts with the definition of the most specific kinds, the leaves of the hierarchy, with subsequent grouping of these kinds into more general concepts. For example, one can start by defining kinds for thermal analysis and electric analysis. He/she then creates a common super class for these two kinds "Stress analysis" which in turn is a sub kind of part analysis.

- A combination development process is a combination of the top-down and bottom-up approaches: One can define the more salient concepts first and then generalize and specialize them appropriately He/she might start with a few top-level concepts and a few specific concepts and then relate them to a middle-level concept.

iii.   Identify Proto Relations

A proto relation is the result of a preliminary attempt at individuating a relation. Proto relations express hypothesized associations between proto kinds or between kinds. The identification and characterization of relations is often the most difficult part of knowledge capture. The identification of proto relations refers to the activity of recognizing the existence of, or becoming attuned to, a particular proto relation in the domain. Characterization follows identification, and refers to the activity of identifying and specifying the properties of a proto relation in a manner that will allow the relational knowledge to be used for making useful inferences at some future time. Thus, recognizing that a tool post is "Above" the lathe bed is the act of discovering and asserting its existence and giving it a name. Characterizing it will involve making assertions such as: the above relation is transitive. Suppose consider the relation between a part to be designed and the different kinds of equipment in the design process. Design drawing tool (such as Auto CAD software) typically require information about the detailed geometry of a part in order to draw the part. However, a decision analysis tool does not require more information than the alternatives. At the same time, the material storage devices require "minimum enclosing box" dimensions and the part weight information to perform the storage function. These associations are now treated as proto relations. The Relation Schematic shown in the following figure facilitates the conceptual analysis of this proto relation.

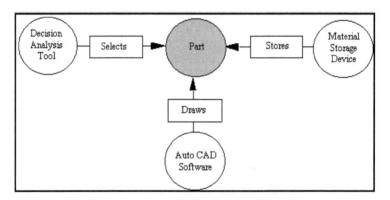

Figure B.4. Part Equipment Relationship

Step 6. Refine and Validate Ontology: This activity will refine and validate the ontology to complete the development process. The refinement process is essentially a deductive validation procedure: the ontological structures are tested with actual data, and the result of the instantiation is compared with the ontology structure. If the comparison produces mismatches, every such mismatch must be adequately resolved. Ontology refinement includes the following steps.

i.   *Kind Refinement Procedure*

The kind refinement procedure is summarized in the following steps:

a.   Make instances of the kinds (and proto kinds). The examples may be constructed from the available source data (source data catalog), otherwise new data must be gathered for the purpose of constructing these examples. The examples must be reasonably representative, with at least one exception case included, if possible. Each of the (proto) kind instances created is populated with properties. Classification diagrams and kind specification forms are used to support the kind instantiation process.

b.   Record information that cannot be recorded in the kind instances. Determine whether this additional information is really necessary, and if so refine the structure of the kind to include the information.

c.   Check whether two instances of the same kind have different defining properties. In such cases, check whether the viewpoints are different. If

267

not, the inconsistencies will have to be resolved by refining the ontology (for instance, by redefining the contentious property to be non-defining).

ii. *Relation Refinement Procedure*

The relation refinement procedure is summarized in the following steps:

a. Make instances of the relations (and proto relations). The examples may be constructed from the available source data (source data catalog), otherwise new data must be gathered for this purpose. The ontology relation diagrams and the relation specification forms are used to aid the instantiation and validation procedure.

b. The properties of each of the relation instances are compared with the properties identified in the ontology description, and any mismatches are resolved. Moreover, check for missing relation properties, and add them if needed.

c. Sample instances of selected relations. Check whether two or more instances of such relations are incompatible. For example, one relation says that a fastener must have a sealant and another may say that it cannot have a sealant. Such inconsistencies may be either due to hidden viewpoint differences not recorded in the ontology, or because of differing viewpoints. Incompatibilities that occur because of differing viewpoints may be resolved by splitting the focus relation into different relations, one for each viewpoint. Otherwise, a consensus must be reached to resolve the incompatibility through discussions with the domain expert.

    d. Detect new relations discovered by example that are not captured in the ontology. Add such relations to the ontology.

Step 7.   Check consistency & accuracy of ontology: This activity will ensure the accuracy and consistency of captured knowledge. Once ontology building is complete, it is necessary to check the consistency and accuracy of information captured specially for design ontology. This consistency check can be done by the help of consistency matrix, which was developed earlier. The detail steps of using the consistency matrix are discussed as follows. If consistency check comes acceptable, the ontology will be ready to use.

This methodology is straightforward and easy to use for manufacturing enterprises. It has the following steps.

    i.   Identify the design activities with relative importance of the domain of interest

    ii.   Prioritize design activities of the domain

    iii. Check captured design activities against the domain design activities

    iv. Calculate consistency index

Detail descriptions of these steps are discussed in the following sections.

Step i. Identify all possible design activities with the relative importance in the related domain of product and process design. If one is building design ontology for the design of a sports utility vehicle, he/she needs to find the design activities in the domain of automobile design. Identifying domain design activities can be done using the following steps.

a. Check for available domain ontology for that particular domain or similar domain and use them appropriately.

b. Search the literatures, best design practices which, includes journal articles, books, conference proceedings, web sites, etc for the design activities.

c. Interview related domain experts or design specialists associated with products and process design.

d. Attend conferences, symposiums, industry group discussions, etc related to product and process design.

Step ii. Prioritize the design activities found in step i. This prioritization is on the basis of relative importance of the design activities in the design process. This provides an important glimpse to the ontology author about what to include in the ontology and what is not. Prioritization can be done using ABC analysis or Pareto analysis. Pareto analysis (sometimes referred to as the 80/20 rule and as ABC analysis) is a method of classifying items, events, or activities according to their relative importance (Balling, Richards, 2000). It is frequently used in inventory management where it is used to classify stock items into groups based on the total annual expenditure for, or total stockholding cost of, each item. But it can be used in situation where prioritization is the main task. Organizations can concentrate more detailed attention on the high value/important activities. A Pareto analysis for prioritizing design activities consists of the following steps.

a. List all design activities of the domain of interest

b. Enter relative importance of each activity

c. Calculate the percentage of total importance represented by each activity.

d. Rearrange the list. Rank items in descending order by total value, starting at the top with the highest value.

e. Calculate the cumulative percentage of the total value for each item at the top; add the percentage to that of the item below in the list.

f. Choose cut off points for A, B and C categories.

g. Present the result graphically. Plot the percentage of total cumulative value on the Y-axis and the item number in X-axis. A typical Pareto curve for a design situation is shown in the following figure.

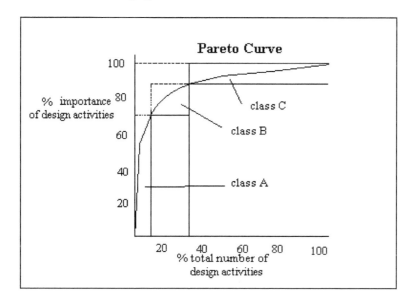

Figure B.5. Pareto Curve of Design Activities

271

Step iii. Check captured design activities against the domain design activities found in step i and ii. This can be done with the help of a matrix shown in the following figure. The columns of this matrix represent the type A, type B, and type C activities of the domain and the rows of the matrix represent the captured design activities of design ontology, which are under construction.

Captured Design Activities in the Design Ontology		Design Activities																											
		Type A					Type B								Type C														
		Activity # 1	Activity # 2	Activity # 3	Activity # 4	Activity # 5	Activity # 6	Activity # 7	Activity # 8	Activity # 9	Activity # 1	Activity # 1	Activity # 1	Activity # 1	Activity # 1	Activity # 1	Activity # 1	Activity # 1	Activity # 1	Activity # 1	Activity # 2	Activity # 2	Activity # 2	Activity # 2	Activity # 2	Activity # 2	Activity # 2	Activity # 2	Activity # 2
Industrial	Activity # 1																												
	Activity # 2																												
	Activity # 3																												
	Activity # 4																												
	Activity # 5																												
Conceptual	Activity # 6																												
	Activity # 7																												
	Activity # 8																												
	Activity # 9																												
	Activity # 10																												
Embodiment	Activity # 11																												
	Activity # 12																												
	Activity # 13																												
	Activity # 14																												
	Activity # 15																												
	Activity # 16																												
Detail	Activity # 17																												
	Activity # 18																												
	Activity # 19																												
	Activity # 20																												
	Activity # 21																												
	Activity # 22																												
Process	Activity # 23																												
	Activity # 24																												
	Activity # 25																												
	Activity # 26																												
	Activity # 27																												
	Activity # 28																												
Raw Score																													

Figure B.6. Consistency Check Matrix

This matrix counts the number of checks in each area of type A, type B, and type C. Bottom line of the matrix gives the sum of theses counts under each category. This raw score is used to calculate the index in the next step.

Step iv. Calculate the consistency index. This index provides a clear picture of the accuracy and consistency of design information, which is captured to construct the ontology. Index calculation can be done using the following steps.

a.   Normalize the raw score. From the consistency cheek matrix in step 3, three different raw scores are found; they are raw score for type A activities (R_A), raw score for type B activities (R_B), and raw score for type C activities (R_C). Normalization of these scores can be done using the following equations.

$$N_A = {R_A}/{M_A} \; X100 \; ;$$ where N_A represents the normalized score for type A

activities and M_A represents the maximum raw score for type A activities.

$$N_B = {R_B}/{M_B} \; X100 \; ;$$ where N_B represents the normalized score for type B

activities and M_B represents the maximum raw score for type B activities.

$$N_C = {R_C}/{M_C} \; X100 \; ;$$ where N_C represents the normalized score for type C

activities and M_C represents the maximum raw score for type C activities.

b.   Calculate the index. Pareto analysis is based on Pareto principle, which says that typically, type A accounts the first 20% of activities in the list will account for approximately 70% of cumulative importance. The next type B accounts 30% of activities, will, typically, account for a further 15% of cumulative importance. These can be subject to less precise control methods. The last type C accounts for the rest 50% of (low importance) activities then account for a mere 15% of importance and can be controlled with a simple system. From this basic principle, the relative weight of each type of activities can be presented as W_A =0.7, W_B =0.15, and W_A =0.15, where W_A, W_B, and W_C are the relative weight of type A, type B, and type C activities respectively. Hence the Consistency Index (CI) of design ontology will be as follows.

$$CI = W_A N_A + W_B N_B + W_C N_C$$

$$\text{or } CI = W_A \frac{R_A'}{M_A} X100 + W_B \frac{R_B'}{M_B} X100 + W_C \frac{R_C'}{M_C} X100$$

The result of this index will be a numeric value between 0 and 100. The higher the value, the more consistent the ontology is. If the value of this index is low for a design ontology, there is question of validity of such ontology model.

Step 8. Collect additional data and analyze data: If consistency check comes unacceptable, further data collection and analysis will be conducted to resolve the disputes. On the basis of this additional data initial ontology will be developed and refinement will be conducted. Finally consistency will be checked again until it meets the acceptable consistency.

Step 9. Incorporate lessons learned and publish ontology: This will add new findings and new research in the ontology and make available for others to use. Ontology must be dynamic and updated in terms of information content. It must be capable to incorporate new findings and lesions learned and publish in the online ontology repository for others.

Although these steps of design methodology are listed sequentially, there is a significant amount of overlap and iteration between the activities. Thus, for instance, the initial ontology development (Step # 5) often requires the capture of additional data and further analysis (Step # 4). Each of the nine activities will involve other activities and tasks.

GLOSSARY OF TERMS

Actor- A design expert, collaborative design team, or other entity capable of actively participating in an activity.

Analysis Of Design Need- A process of identifying and clarifying product's functional requirements in detail level.

Analyze & Refine Configuration- A process of determining configuration alternatives considering all aspect of design principles.

Analyze, Refine & Evaluate Design Variables- A process of identifying feasible & optimal values design variables and determine their expected performance.

Anticipatory Failure Determination (AFD)- Is a failure analysis method. Like FMEA, it has the objective of identifying and mitigating failures. Rather than asking developers to look for a cause of a failure mode, it reverses the problem by asking developers to view the failure of interest as the intended consequence and try to devise ways to assure that the failure always happens reliably.

Assembly Drawings- Drawings that show the components that makes up a product or subassembly and the location of each part in relation to the other parts.

Axiomatic Design- Recognizes four domains. The needs of the customer are identified in customer domain and are stated in the form of required functionality of a product in functional domain. Design parameters that satisfy the functional requirements are defined in physical domain, and, in process domain, manufacturing variables define how the product will be produced.

Backtracking- Redoing some phase of the design process.

Benchmark- A standard by which something can be compared; a product used to compare a design.

Bending- A manufacturing process that plastically deforms sheet metal using a matched punch-and-die set, or a descending punch to wipe-form the work piece over the edge of the die.

Beta Prototype- A full-scale, functional part or product prototype using materials and manufacturing processes that will be used in production.

Beta Testing- Is the testing a nearly finished version of a piece of software or hardware, with the goal of finding defects missed by the developers. Often beta testing is carried out by people outside of the developers' organization.

Bill Of Materials (BOM)- A table of product component information organized with column headings for part number, part name, material, quantity used in assembly, and other special notes.

Blanking- A sheet metalworking process that shears a smaller shaped piece of sheet metal, called a blank, from the stock sheet; blanks are later used in deep drawing.

Blow Molding- A process used to form polymers wherein a molten material is injected with air then expands to the shape of the mold.

Boring- A machining process that increases the diameter of an existing hole by feeding a sharp tool into the rotating work piece.

Brazing- A process used to join two metal pieces together with the addition molten brass or zinc solder.

Budgeted Cost Of Work Performed (BCWP)- The sum of the products of the percent complete for each work task times the amount budgeted each work task.

Built-in-Self-Test - A feature of automatic testing where many test pattern programs are built directly into the circuit generally for go/no-go testing of the assembly or circuit using signature analysis.

Bulk Deformation- Manufacturing processes that change the shape or form of bulk raw materials by compressive or tensile yielding.

Casting- Processes in which molten metal is poured into a cast to solidify; used to produce complex geometries within broad tolerances.

Clearance Fit- An intentional space between mating parts such as between a bolt and hole.

Coefficient Of Friction- A relative measure of the amount of friction force between two surfaces; equal to the ratio of the friction force divided by the force normal to the surface.

Coefficient Of Thermal Expansion- A measure of the amount a material elongates in response to a change in its temperature.

Collaborative Design Team- Is a team consisting of representatives from marketing, engineering, manufacturing, finance, purchasing, test, quality, finance and any other required disciplines with responsibility for developing a product or product subsystem. This team is empowered to represent the functional disciplines and develop a product by addressing its life cycle requirements including its product and support.

Company Requirements- Specifications for the design of a product that originate from the company rather than the customer. Company requirements include marketing, manufacturing, financial, and legal considerations.

Component Decomposition- The process of identifying and separating the components of a product into parts and subassemblies.

Compression Molding- A process that forms a charge of thermoset or elastomer between heated mold halves under pressure while the material cures.

Computer-Aided Design (CAD)- Is the use of a computer to assist in the creation and modification of a design, most commonly, designs with a heavy engineering content.

Computer-Aided Engineering (CAE)- Is the use of computers in design, analysis, and manufacturing of a product, process, or project. Sometimes refers more narrowly to the use of computers only in the analysis stage.

Computer-Aided Process Planning (CAPP)- Computer software that establishes the type and sequence of manufacturing processes for each part produced.

Concept- Is an idea for a new product or system that is represented in the form of a written description, a sketch, block diagram or simple model. A concept is the earliest representation of a new product or of alternative approaches to designing a new product.

Conceptual Design- An abstract embodiment of a working principle, geometry, and material; a phase of design when the physical principles are selected.

Concept Evaluation- A process of evaluating all generated alternatives to find out the feasible and producible alternatives and predict their performances.

Concept Generation- A process of creating alternatives or design concept, which is the earliest representation of a new product or of alternative approaches to designing a new product.

Concept Selection- A process of selecting best alternatives among available design alternatives.

Concept Testing- The process by which a concept statement, sketch or model is presented to customers for their reactions. These reactions can either be used to permit the developer to estimate the sales value of the concept or to make changes to the concept to enhance its potential sales value.

Configuration Design- The selection and arrangement of features on a part; or the selection and arrangement of components on a product; a phase of design when geometric features are arranged and connected on a part, or standard components or types are selected for the architecture.

Configuration Requirements Sketch- A sketch drawn to approximate scale showing the essential surroundings of a part, including forces, flows, features of mating parts, support points or areas, adjacent parts, and obstructions or forbidden areas.

Configure Parts- A process of determining the number & type of geometric features, relative dimensions, and their arrangements.

Configure Product- A process of determining the number & type of components, their specific functions, and their arrangements.

Consensus Decision- Decision in which the team thoughtfully examines all of the issues, and agrees upon a course of action, which does not compromise any strong convictions of a team member.

Contingency Design- Is a form of mistake-proofing focusing on the user's experience with the product. The intent is to design in features that help the user avoid mistakes or allow the users to quickly correct input of data or operation of the product. This is accomplished through layout and graphic design, intuitive operation, clear instructions, appropriate markings and warnings, descriptive error messages, avoidance of technical jargon, and simple operation steps.

Correlation Ratings Matrix- A portion of the house of quality listing values that rate the correlation between qualitative customer requirements and quantitative engineering characteristics.

Customer Requirements- Specifications for the design of a product that originate from the customer rather than the company. Customer requirements include functional performance, operating environment, human factors, safety, robust ness, maintenance, and repair.

Customer Requirements Analysis- A process of analyzing customer requirements to check the feasibility of the design and identify design requirements.

Decision-Making Process- The process used to identify alternatives and the outcomes associated with each alternative, and to judge the outcomes to make a selection.

Design- A set of decision-making processes used to determine the form of a product given the functions desired by the customer; processes to prescribe the sizes, shapes, material compositions, and arrangements of parts so that the resulting machine will perform a required task; a package of information such as and specifications sufficient to manufacture a product.

Design Activity- Any process of transforming state of the environment that happens over time.

Design Analysis- A portion of the design process that predicts or simulates performance of each alternative, reiterating to assure that all the candidates feasible.

Design Analysis & Trade Off- A phase of design that predicts or simulates performance of each alternative, reiterating to assure that all the candidates are feasible.

Design Concept- The abstract embodiment of a physical principle, material, geometry; same as concept design.

Design Evaluation- The portion of the design process that assesses or weighs from the analyses to determine which alternative is the best.

Design Expert- A human being entity that performs activities to achieve design goals. Its ability to actively participate in the planning process is what separates a design expert from a machine.

Design For Adjustability- An approach used to design products and/or machinery by designing adjustable parts to fit all human beings such as a microphone stand, or car seat.

Design For Assembly (DFA)- Refers to the principles of designing assemblies so that they are more manufacturable. DFA principles address general part size and geometry for handling and orientation, features to facilitate insertion, assembly orientation for part insertion and fastening, fastening principles, etc. The objective of DFA is to reduce manufacturing effort and cost related to assembly processes.

Design For Close Fit- A design approach that establishes sets of sizes to match classes of customers, such as shoe sizes, hat sizes, or ring sizes.

Design For Cost- A development methodology that treats cost as an independent design parameter. A realistic cost objective is established based on customer affordability, tradeoffs are made between the cost objective and other product functions/parameters, cost models are used to project the cost early in the development cycle, and a variety of techniques such as function analysis and DFM are used to proactively achieve the cost objective.

Design For Disassembly (DFD)- Is a set of principles used to guide designers in designing products that are easy to disassemble for recycling, remanufacturing, or servicing.

Design For Lifecycle Cost- This represents the totality of design-to-cost addressing all costs related to acquisition, operation, support and disposal.

Design For Manufacture- Design to maximize ease of manufacture by simplifying the design through part-count reduction, developing modular designs, minimizing part variation, designing a part to be multi-functional, etc.

Design For Manufacturability- Is a methodology for designing product's in a way that facilitates the fabrication of the product's components and their assembly into the overall product. In this respect it is synonymous with Design for Manufacturability / Assembly.

Design For Performance- Designed to perform to product requirements under a wide variety of manufacturing and user operating conditions. Without this there may be no product, so be sure that the requirements are really what customers need.

Design For Reliability- Designing the product so it works the first time, every time for the life of the product (decreasing cycle failure).

Design For Safety- Design so that the manufacture of and the use or abuse of the product minimize the possibility of injuries which could lead to product liability problems. There are Federal requirements to be met. DFS experts in your company or as consultants know the rules and many opportunities. Designers should use DFS Checklists and published signage and labeling standards.

Design For Serviceability -Design to ease maintenance and service required during the life cycle of the products. The design of the support processes needs to be developed in parallel with the design of the product. This can lead to lower overall life cycle costs and a product design that is optimized to its support processes. It is a set of principles and a methodology for analyzing product concepts or designs for characteristics and design features that reduce service requirements and frequency, facilitate diagnosis, and minimize the time and effort to disassemble, repair/replace, and reassemble the product as part of the service process.

Design For Testability / Inspectability- It facilitates testing of a product in the design process. Increasingly complex and sophisticated products require capabilities and features to facilitate test and acceptance of products and diagnosis products if a defect is identified.

Design For The Environment- A group of design methods that aim to minimize the use of raw materials, increase recyclables, improve remanufacture, increase energy efficiency, and improve the workplace environment.

Design For The Extreme- Design method that strives to fit the smallest or largest person.

Design For *X*- A term used to describe any of the various design method that focus on specific product development concerns.

Design Method- A procedure or set of guidelines for solving a design problem.

Design Object- Any tangible or conceptual entity in the design domain.

Design Phase- A period or stage in the design of a part or product; concept design phase, configuration design phase, parametric design phase.

Design Problems- Product deficiencies that require resolution; product Opportunities that require consideration.

Design Process- The problem-solving process used to formulate a design problem, generate alternative solutions, analyze and evaluate the feasibility and performance of each alternative, and select the best alternative, reiterating if necessary.

Design-Project Report- A report that summarizes the work tasks undertaken in a design project and discusses the recommended design in detail; usually includes content on the nature of the design problem, design formulation, concept design, configuration design, parametric design, prototype testing, and detail design description and performance.

Design Reviews- Are formal technical reviews conducted during the development of a product to assure that the requirements, concept, product or process satisfies the requirements of that stage of development, the design is sound, the issues are understood, the risks are being managed, any problems are identified, and needed solutions proposed. Typical design reviews include: requirements review, concept/preliminary design review, final design review, and a production readiness/launch review.

Design Tool- A non-human entity, which has the capacity to carry out design functions.

Design Variable- A parameter that can be arbitrarily selected by the designer that influences the behavior of the design candidates; a controllable variable.

Design Validation- Testing to assure that the product conforms to defined user needs and requirements. This normally occurs toward the end of the Design Phase following successful design verification and prior to pilot production, beta/market testing, and product launch. Design validation is normally performed on the final product under defined, operating conditions. Multiple validations may be performed if there are different intended uses.

Design Verification- Is the process of ensuring the design conforms to specification (design outputs meet design input requirements). Design verification may include: alternate calculations, design reviews, comparison to similar designs, inspection, and system or product testing.

Detail Design- The conversion of product specifications into designs and their associated process and/or code-to documentation. Detailed design includes design capture, modeling, analysis, developmental testing, documentation, process design, producibility analysis, test plan development, coding, and design verification and validation.

Detail Drawings- Drawings showing orthographic projection views of a part (e.g. front, side, or top); showing geometric features drawn to scale along with full dimensions, tolerances, manufacturing process notes, and title block; detail drawings sometimes include section views and oblique plane views.

Detail Functionality Analysis- A phase of design that determines the functional characteristics of each component using extensive analysis.

Detail Systems Design- A phase of design that determines the required systems specifications for the product to be produced.

Diagram- Illustration that is intended to explain how something works or the relationship between the parts.

Die Casting- The solidification of molten material after it is injected into a mold under high pressure.

Drawing- The plastic deformation of a bar, rod, or wire as it is pulled through successively smaller dies; also, the plastic deformation of sheet metal into a die, forming cupped, box, or hollow parts.

Drilling- A machining process that removes material from the work piece using a rotating bit, thus forming a hole.

Early Supplier Involvement- Is the process of getting the supplier involved early in the development process (when an item is being conceptualized, designed or specified) so that the supplier can make proactive suggestions to improve the design and reduce its cost vs. providing reactive feedback once the design has been completed.

Earned Value Analysis- A method used in project engineering to determine, a project is ahead or behind schedule and either over or under budget.

Electrical Discharge Machining- A process of removing metal by means of an electrical discharge spark.

Electronic Design Automation (EDA)- Consists of hardware and software tools to aid in the design and development of electronic products through design capture, simulation, synthesis, verification, analysis, and testing.

Embodiment- Manifestation or incarnation of something.

Embodiment Design- A process of selecting detail specifications of product and it's part with the consideration of various aspect of design such as safety, manufacturability, assembly, environment, etc.

Embossing- Plastic indentation of surface to form ribs, beads, or lettering on the surface of metal.

Engineering Analysis- Using tools such as analytical models or empirical equations to predict the performance or behavior of an object or system.

Engineering Change- A modification to a component, product configuration, or document from currently defined and approved status. Changes cause version or revision levels of affected items to be updated.

Engineering Design- The set of decision-making processes and activities used to determine the form of an object given the functions desired by the customer.

Engineering Design Specification (EDS)- Document containing a comprehensive description of intended uses and functional requirements of a product.

Equipment- Machines or apparatus designed to perform a thermal, chemical, or , mechanical process or portion thereof.

Evaluate & Select Best Configuration- A process of selecting best configuration after evaluating all available configuration alternatives.

Evaluation- The process of rating or assessing the predicted perfornlance of feasible design candidates against established evaluation criteria; used to determine the "best" design alternative.

Evaluation Criteria- Standards by which alternatives may be rated, assessed, or ranked.

Extrusion- Bulk deformation process that squeezes heated metal or plastic materials through die producing constant cross section.

Facing- A finishing operation that removes material from a cylindrical surface; the cutting tool moves radially as the work piece rotates.

Fail-safe Design Principle- A principle used to design components such that upon failure of a component, some critical functions are still performed.

Failure Analysis- is a collection of techniques to determine the root cause of a component or process defect or failure.

Failure Mode And Effects Analysis (FMEA)- Is a procedure in which each potential failure mode in every sub-item of an item is analyzed to determine its effect on other sub-items and on the required function of the item. It is used to identify potential failure modes and their associated causes/mechanisms, consider risks of these failure modes, and identify mitigating actions to reduce the probability or impact of the failure.

Fault Tree Analysis- Is a top-down, hierarchical analysis of faults to identify the various fault mechanisms and their cause. It graphically describes the cause and effect relationships that result in major failures.

Feasible Design- A design candidate that meets design specifications and/or satisfies design constraints.

Field Testing- Is the testing a product in the actual context in which it will be used, as opposed to laboratory testing, or testing the product in its operating environment.

Final Design Reviews & Verification- A phase of design that identifies technical performance, risks, improvement of performance & process, and verifies design parameters.

Finishing- Preparing the final surface of a part to protect it from the environment or to enhance the Visual appearance; e.g.; anodizing, chrome plating, and painting.

Finite Element Analysis (FEA)- A computer-based method that breaks geometry into elements and links a series of equations to each, which are then solved simultaneously to evaluate the behavior of the entire system. Most often it used for structural analysis, but widely applicable for other types of analysis and simulation, including thermal, fluid, and electromagnetic.

Fixture- Tooling designed to locate and hold components in position.

Forging- A deformation process that creates a desired shape by plastically pressing material in two halves of a die set by hammer strokes or a hydraulic press.

Forming- An early stage in team development when participants transition from being an individual to being a team member, politely interacting, learning the nature of the tasks to be performed, the goals of the project, and the personalities and work styles of fellow members.

Formulation- A phase of design in which customer and company requirements are determined, engineering design specifications are prepared, and a solution plan is prepared.

Function Decomposition- Product function decomposition is a process of identifying and separating required sub functions without regard to possible embodiments.

Functional Test- Is a test that identifies functional level faults in printed circuit board assemblies (PCBAs), including manufacturing related faults not identified by in-circuit tests (ICT), timing related failures, and faults internal to components. Functional test equipment operates at the same frequency the PCBA is designed for and may have the capability to margin temperature, voltage and frequency.

Fused Deposition Modeling (FDM)- A rapid prototyping process that deposits a thin filament of melted (fused) material in precise locations on a horizontal layer, using numerically controlled positions; as the material solidifies, the prototype model is built, layer by layer

Generating (alternatives)- The set of activities and decision-making processes used to create alternatives for later analysis and evaluation.

Grinding- Machining process to remove material from a surface with an abrasive wheel or tool.

Hazard Analysis-Is the detailed examination of a product from the user perspective to detect potential design flaws (possibilities of failure that could cause harm) and to enable manufacturers to correct them before a product is released for use.

Highly Accelerated Life Test (HALT)- Is a process developed to uncover design defects and weaknesses in electronic and mechanical assemblies using a vibration system combined with rapid high and low temperature changes. The purpose of HALT is to optimize product reliability by identifying the functional and destructive limits of a product. HALT addresses reliability issues at an early stage in product development.

Highly Accelerated Stress Screening (HASS)- Is a technique for production screening that rapidly exposes process or production flaws in products. Its purpose is to expose a product to optimized production screens without affecting product reliability. Unlike HALT, HASS uses nondestructive stresses of extreme temperatures and temperature change rates with vibration.

Honing- To sharpen or smooth.

Human Factors- Refers to the characteristics of human beings that are applicable to the design of systems and devices of all kinds. It furthers serious consideration of knowledge about the assignment of appropriate functions for humans and machines, whether people serve as operators, maintainers, or users in the system.

Identify Design Tasks- A process of identifying and organizing design tasks, which are necessary to conduct the conceptual deigns process.

Information- Any object that can be used, created, or modified by an activity and whose essence is to act as information in the design domain.

Industrial Design- Is the design that is done in companies and consultancies by people trained in industrial design, or in art and design schools in general. Industrial design focuses on the physical form and interactive properties as opposed to the functioning of the product or system.

Industrial Design- Decisions or activities to determine essential product aesthetics and basic functions.

Industrial Design & Formulation- A process of transforming customer/design requirements into design requirements using design resources and satisfying design constraints. This activity ensures feasibility of the design.

Injection Molding- Process that melts thermoplastic pellets and injects the molten material into a mold under high pressure.

Insertion- Moving a part into another in the assembly for fastening or joining.

Investment Casting- A casting process that uses wax patterns dipped in a slurry to create a mold; wax is removed by melting; lost-wax process.

Lapping- A finishing process used to polish a surface with a slurry of fine abrasive particles.

Machine- Combination of resistant bodies, so arranged that by their means, the mechanical forces of nature can be compelled to do work accompanied by certain determinate motions.

Machine- A non-human entity, which has the capacity to carry out design functions.

Machine Tool- A machine that makes other parts or products.

Machining- A subtractive process that removes material from a workspace by a sharp cutting tool that shears away chips of material, to create desired form or features.

Manufacturing Design-The characteristic of a product's design that facilitates the fabrication of the product's components and their assembly into the overall product.

Material- Any material object that can be used, created, or modified or consumed by an activity in the design domain (e.g. Raw materials, Subassemblies).

Mechanical Press- A machine used for sheet metal working anti forging operations; an electric motor energizes a flywheel that strokes the hammer or punch.

Milling- A machining process that removes material from a flat surface to form slots, pockets, recesses; a cutting tool rotates as the material is fed.

Mistake-Proofing- Improving product designs, tooling designs, or processes to prevent mistakes from being made or to quickly and easily detect or mitigate the effect of a mistake. Mistake proofing involves six principles: elimination, replacement, prevention, facilitation, detection, and mitigation. Also known as error proofing and poka-yoke.

Modular Design- Consists of combining standardized building blocks or "modules" in a variety of ways to create unique finished products. Thus, even though the parts and assemblies may be standardized, the finished product is unique.

NC/CNC Machining -Numerically controlled/computer numerically controlled machining.

New Product Development- Is the business process for developing new hardware, software and service products for the enterprise. It includes all activities from development of the idea or concept for the product, the development of the product and its processes, and the launch of the product into production and into the market place.

Original Design- Development of a new component, assembly, or process that had not existed before.

Painting- Finishing process used to protect a surface or enhance visual appearance.

Parametric Design- A phase of design that determines specific values for the design variables.

Participatory Design- Refers a democratic approach to design that encourages participation in the design process by a wide variety of stakeholders, such as: designers, developers, management, users, customers, salespeople, distributors, etc. The approach stresses making users not simply the subjects of user testing, but actually empowering them to be a part of the design and decision-making process. This is accomplished through direct involvement with the product development team on major projects for one or a small number or customers or through frequent customer or user review and feedback during the development process using mechanisms such as focus groups, web-based customer participation, usability studies, etc.

Planning- Machining process that removes material using a translating cutter as the work piece feeds.

Plating- Finishing process that chemically alters the surface of a part.

Polishing- Finishing process that uses abrasive powders embedded in a rotating leather or felt wheel to remove surface irregularities.

Preliminary Design- Early phases of design, including concept, configuration, and parametric design activities.

Pre-production Prototype- A full-scale part or product made and assembled with , final materials and production-like processes.

Primary Manufacturing Process- The manufacturing process that principally alters the material's shape or form.

Product- Any passive object that can be used, created, or modified by an activity in the design domain. These are typically the inputs and outputs of a design process.

Product Design - A process of transforming customer/design requirements into detail design specification using design resources and satisfying design constraints. This activity happens over time.

Production Process Planning- A process of planning and determining process specifications to produce the designed products.

Process Development- Defining and developing a manufacturing process to accommodate the specific requirements of a given product while, meeting process quality and cost objectives.

Process Design- Defining and designing a manufacturing process to accommodate the specific requirements of a given product while, meeting process quality and cost objectives.

Process Planning- Is the analysis and design of the sequence of processes, resources requirements needed to produce products into workable instructions for manufacture. It also includes the specification and selection of tools, fixtures, equipment and inspection/test requirements.

Product Requirements- A technical characteristic of the product expressed in the developer's language to respond to a customer need. A good requirement should be 1) stated so that it is directly actionable by engineering, 2) is global and does pre-suppose a particular technical solution, and 3) is measurable so that it can be ultimately verified. The developer uses the product requirements to guide the design and building of the product.

Product- Designed object or artifact that is purchased and used as a unit.

Product Development- The portion of the product realization process that begins with formulation activities and concludes with production planning and manufacturing engineering activities; excludes activities beginning with production.

Product Life Cycle- The cycle of birth and death of a product; starts with the introduction of the product into the market, includes stages of growth, maturity, and decline; ends with removal of product from market.

Product Concept Test- A market research activity that uses a reduced-scale or full-scale model of a new product or "product concept"; usually nonfunctional but looks like a "finished" product.

Product Realization Process (PRP)- Design and manufacturing processes that convert information, materials, and energy into a finished product.

Production Design- Designing and planning the type and arrangement of equipment and the use of labor in a factory to make a product.

Production Readiness- Review is a design review conducted prior to putting a product into production. This reviews assesses whether all needed product and process data has been completely generated, that the production process has been validated, and that the company is ready to begin production (either pilot production, low-rate initial production, production ramp-up, or full-rate production).

Project Control- Detecting whether a project is on time and within budgeting upon and implementing corrective actions if necessary.

Project Plan- A package of key items to be completed in a project including problem statement, mission statement, project objectives, work breakdown structure.

Prototypes- A physical model or representation of the new product concept or design. Depending upon the purpose, prototypes may be non-working models or representations, functionally working, or both functionally and geometrically complete and accurate. Prototypes (physical, electronic, digital, analytical, etc.) can be used for the purpose of, but not limited to: a) assessing the feasibility of a new or unfamiliar technology, b) assessing or mitigating technical risk, c) validating requirements, d) demonstrating critical features, e) qualifying a product, f) qualifying a process, g) characterizing performance or product features, or h) elucidating physical principles.

Punching- Sheet metal working processes that produces features such as slots, notches, extruded holes, and holes using a punch and die.

Rapid Prototype-More generally, it is the process of quickly generating prototypes or mockups of what a product system will look like. Rapid prototyping may be done with paper prototypes such as sketches, low-fidelity physical prototypes, CAD visualization, rapid application development, or video prototyping.

Reaming- Refining the diameter of an existing hole.

Redesign- The process of revising any portion of an existing product's form (i.e., shape, configuration, size, materials, or manufacturing processes); selecting new values for the design variables, reanalyzing, and reevaluating, to obtain better performance and improve customer satisfaction.

Reliability- A sub-set of statistical engineering methodology, which predicts performance of a product over its intended life cycle and understanding of the effects of various failure modes on system performance.

Reliability Analysis- A predictive tool used to estimate the "life" of a product. This is usually expressed in terms of hours as "mean time between failure" (MTBF).

Redundant Design Principle- A principle used to design components such that additional components or systems, configured in parallel or series, take over the , "principle function of the failed component or system.

Requirements Analysis-The determination of product-specific performance and functional characteristics based on analyses of: customer needs, expectations, and constraints; operational concept; projected utilization environments for people, products, and processes; and measures of effectiveness.

Reviews & Feasibility Analysis- A process of reviewing designs requirements to make sure that the design is feasible and manufacturable using design resources. This activity identifies engineering design needs.

Risk Assessment- A consideration of the likely risks that face a project and possible contingency plans to overcome such situations.

Robust Design- Methods used to design robust products, which is one that performs in spite of variations in its material properties, how it was manufactured, the operating environment, or how it is used.

Robust Design- Design of the product in a manner to desensitize the product to variation including misuse and increase the probability that it will perform as intended.

Robust Design- In its most general sense insures operation in a variety of environments, throughout life. Environmental Stress Testing weeds out problems by subjecting samples to a simultaneous set of extreme operating conditions.

Rolling- Bulk deformation process used to form sheets, bars, rods, and structural shapes by plastically compressing slabs, billets, and blooms between two rollers.

Safe-Life Design Principle- A design principle used to design components to operate for their entire predicted useful life without breakdown or malfunction.

Sand Casting- Molten metal solidifies in a mold made of sand. Standard mold is formed by packing sand around a pattern with the same external shape as the part of the cast.

Sawing- Cutting/dividing material with a toothed blade.

Secondary Manufacturing Process- Manufacturing process that adds or removes geometrical features from the basic forms.

Select Design Variables- A process of identifying design variables and determine values of those design variables.

Sectioned Assembly Drawing- A cutaway portion of the assembly drawing that exposes the details of an interior portion of the assembly.

Selection Design- Decision-making processes used to match the desired functional requirements of a component with the actual performance of standard components listed in vendors' catalogs.

Selective Laser Sintering (SLS)- A rapid prototyping process that uses a high-power laser to sinter together fusible materials, such as powdered metals, layer by layer.

Sensitivity Analysis- Analyzing the contribution of a part's variance to the total variance of the assembly.

Shaping- Machining process that removes material from a translating work piece and a stationary cutter.

Shearing- Cutting or separating sheet metal along a straight line; used to size sheets for subsequent operations.

Sheet Metalworking- Permanent deformation of thin metal sheets produced by bending or shearing forces; often called *stamping*.

Soldering- Process used to join two metal pieces together with the addition of molten tin, lead, and silver alloys.

Solid Modeling- A geometric modeling method that completely and unambiguously describes both the exterior and interior of a part or assembly in three dimensions (geometry, topology and mass properties).

Stamping - Sheet metalworking processes.

Standard Part- A common interchangeable item, having standard features, typically mass-produced and used in various applications; e.g., nut, bolt, screw, washer, lubricant.

Standard (Sub) Assembly- One that is routinely manufactured for general use; e.g., pump, motor, valve, switch.

Subassembly- An assembly that is included in another assembly or subassembly.

Tertiary Manufacturing Process- Surface treatment such as polishing, painting, heat-treating, and joining.

Thermal Forming- Vacuum forming thin sheets of thermoplastic.

Tinkering- Repetitive or iterative cutting and trying, fabrication, and testing; does not use scientific principles or mathematics to predict behavior.

Top-Down Design- Is a design methodology whereby an entire design is decomposed into its major components, and then these components are further decomposed into their major components, etc. The constraints are established early in the design flow, and then are passed on and adhered to by the back-end processes.

Turning- Machining process that removes material from rotating work piece; lathes.

Variant Design- Type of design; modifying the performance of an existing product by varying some of its design variable values or product parameters such as size, or specific material, or manufacturing processes.

Value Engineering- A systematic approach to evaluating design alternatives that seeks to eliminate unnecessary features and functions and to achieve required functions at the lowest possible cost while optimizing manufacturability, quality, and delivery.

Verification And Validation Of Processes- A process of finalizing the production process specifications and release for production.

Virtual Prototype- Non real, electronic prototype; modeled inside the memory of a computer.

Welding- Fastening process that permanently joins two or more metal parts by controlled melting; fusion of metals.

Wire Drawing- Process that transforms bar stock by pulling it through a set of successively narrowing dies, forming a long strand of wire that is usually wound on a spool as a continuous process.

Work Breakdown Structure- Is a hierarchical tree structure decomposing a project into activities and sub-activities to help define and control the project and its elements of work.

Worst-Case Tolerance Design- A method that assumes that each process will produce parts with the "worst" precision within its capability.

BIBLIOGRAPHY

Alsene, Eric, "The Computer Integration of the Enterprise", IEEE Transaction on Engineering Management, vol. 46, n. 1, 1999, pp 26-36.

Amice, C., "Open System Architecture for CIM", CIM-OSA, Springer, Berlin, 1993.

Andersin, H., Reinikka, M. and Wickstrom, LeRoy A., "Enterprise integration- metrics for improvement and benchmarking", Computer Applications in Technology, n B-17, 1994, pp 227-239.

Ashley, Braganza, "Enterprise Integration: Creating competitive capabilities", Integrated Manufacturing Systems, vol. 13, n. 8, 2002, pp 562-572.

Bailetti, Antonio J. John R. Callahan, and Sean McCluskey, "Coordination at different stages of the product design process", R&D Magazines, vol. 28, n. 4, 1998.

Bierly III, Paul E. a, Joseph E. Coombsb, "Equity alliances, stages of product development, and alliance instability", Journal of Engineering Technology Management, Vol. 21, 2004, pp 191–214.

Bill Swartout, Ramesh Patil, Kevin Knight and Tom Russ, "Toward Distributed Use of Large-Scale Ontologies", Proceedings of the Tenth Knowledge Acquisition for Knowledge-Based Systems Workshop, November 9-14, 1996, Banff, Alberta, Canada.

Brosey, William, D., R.E. Neal and D.F. Marks, "Grand Challenges of Enterprise Integration", IEEE Symposium on Emerging Technologies and Factory Automation, vol. 2, 2001, pp 221-227.

Bunge, M. A., Treatise on Basic Philosophy Volume 3: Ontology I - The Furniture of the World, Kluwer Academic Publishers, Dordrecht, 1977.

Calmeta, Ricardo, Christina Campos and Reyes Grangel, "Reference Architectures for Enterprise Integration", The journal of systems and software, vol. 57, 2001, pp 175-191.

Cao, Yan; Liu, Ning; and Zhao, Rujia, " Research on enterprise integration based on virtual manufacturing system", Computer Integrated manufacturing Systems, CIMS, vol. 5, n. 3, 1999, pp 25-29.

Chapman, Ross and Paul Hyland, "Complexity and learning behaviors in product innovation", Technovation, vol. 24, 2004, pp 553–561.

Chen Hui and David R. Shonnard, "Systematic Framework for Environmentally Conscious Chemical Process Design: Early and Detailed Design Stages", Ind. Eng. Chem. Res., 2004, pp 535-552.

Chin, Kwai-Sang and Wong, T. N., "Integrated product concepts development and Evaluation", International Journal of Computer integrated manufacturing, vol. 12, no. 2, 1999, pp 179 –190.

Crilly, Nathan, James Moultrie and P. John Clarkson, "Seeing things: consumer response to the visual domain in product design", Design Studies Vol. 25 No. 6, November 2004, pp 547–577.

Deng, Zigiong, "Three level model for enterprise integration", Proceedings of the Chinese Society of Electrical Engineering, vol. 17, n. 5, 1997, pp 8- 14.

Dixon, J.R., and C. Poli, Engineering Design and Design for Manufacturing, Conway, MA, Field Stone Publisher, 1995.

Don Liles, "Operate a small integrated manufacturing enterprise", The Automation & Robotics Research Institute, University of Texas at Arlington, 1995.

Dragan Djuriæ, Dragan Gaševiæ, Vladan Devedžiæ, "Ontology Modeling and MDA", Journal of Object Technology, Vol. 4, No. 1, 2005.

Eeckhout, Lieven, Bosschere, Koen De, "How accurate should early design stage power/performance tools be? A case study with statistical simulation", The Journal of Systems and Software, vol. 73, 2004, pp 45–62.

Eggert, Rudolph J., Engineering design, Pearson Prentice Hall, New Jersey, 2005.

Green, P. and Rosemann, M., "Integrated Process Modeling. An Ontological Evaluation," Information Systems, vol. 25, n. 2, 2000, pp 73-87.

Green, P. and Rosemann, M., "Perceived Ontological Weaknesses of Process Modeling Techniques: Further Evidence," in Proceedings of the 10th European Conference on Information Systems, S. Wrycza (Ed.), Gdansk, 2002, pp 312-321.

Green, P. and Rosemann, M., "Applying Ontologies to Business and Systems Modeling Techniques and Perspectives: Lessons Learned," Journal of Database Management, vol. 15, n. 2, 2004, pp 105-117.

Green, P., Rosemann, M. and Indulska, M., "Ontological Evaluation of Enterprise Systems Interoperability Using ebXML," IEEE Transactions on Knowledge and Data Engineering, vol. 17, n. 5, 2005, pp 713-725.

Green, P., Rosemann, M., Indulska, M. and Manning, C., "Candidate Interoperability Standards: An Ontological Overlap Analysis," Working Paper Business School, University of Queensland, 2004.

Gruber, T. R., Ontolingua: A Mechanism to Support Portable Ontologies, Knowledge Systems Laboratory Technical Report KSL 91-66, Final Version, Stanford University, 1992.

Gruber, T. R., "A Translation Approach to Portable Ontology Specification", Knowledge Acquisition, vol. 5, 1993, pp 199-220.

Gunasekaran, A., "Agile manufacturing: enablers and an implementation framework", International Journal of Production Research, vol. 36, n. 5, 1998, pp 1223-1247.

Gruninger, M., Atefi, K., and Fox, M.S., "Ontologies to Support Process Integration in Enterprise Engineering", Computational and Mathematical Organization Theory, Vol. 6, No. 4, 2000, pp 381-394.

Harrold, Dave, "Enterprise Integration requires understanding the plant floor", Control Engineering, vol. 47, n. 2, 2000.

Hobbs, J., W. Croft, T. Davies, D. Edwards and K. Laws, The TACITUS Commonsense Knowledge Base, Artificial Intelligence Research Center, SRI International, 1987.

Integrated Manufacturing Technology Road mapping Project – An overview of the IMTR roadmaps, 2004, IMTI, Inc.

310

Jennings, N. R., K. P. Sycara, *et al.,* "A Roadmap of Agent Research and Development", Journal of Autonomous Agents and Multi-Agent Systems, vol. 1, n. 1, 1998, pp 7-36.

Junior, Wilson Kindlein, Andrea Seadi Guanabara, "Methodology for product design based on the study of bionics", Materials and Design, vol. 26, 2005, pp 149–155.

Karsak, E. Ertugrul, "Fuzzy multiple objective decision making approach to prioritize design requirements in quality function deployment", International Journal of Production Research, Vol. 42, no. 18, 2004, pp 3957–3974.

Keen, C. and Lakos, C., "Analysis of the Design Constructs Required in Process Modeling," in Proceedings of the International Conference on Software Engineering: Education and Practice, Dunedin, 1996, pp 434-441.

Kim, H.M., "Representation and Reasoning About Quality using Enterprise Models", PhD Dissertation, Enterprise Integration Laboratory, Department of Mechanical and Industrial Engineering, University of Toronto, 1999.

Kim, H.M., Fox, M.S., and Gruninger, M., "An ontology for quality management - enabling quality problem identification and tracing", BT Technology Journal, Vol. 17, No. 4, 1999, pp 131-140.

Kolb, D., Experiential Learning: Experience as the Source of Learning and Development, Prentice-Hall, 1984.

Kosanke, K., Vernadat, F. and Zelm, M., "CIMOSA: enterprise engineering and integration", Computers in Industry, vol. 40, 1997, pp 83-97.

311

Kumar, P., "Production design formulation and forms", Journal of the Institution of Engineers (India), Vol. 81, No. 2, 2000.

Lim, Soon H., Neal Juster and Pennington, A., "Enterprise Modeling and Integration: a taxonomy of seven key aspects", Computers in Industry, vol. 34, 1997, pp 339-359.

Madu, Christian N. and Madu, Assumpta A., "E- quality in an integrated enterprise", TQM Magazine, vol. 15, n. 3, 2003, pp 127-136.

McGuinness, D.L., Fikes, R., Rice, J. and Wilder, S., "An Environment for Merging and Testing Large Ontologies. Principles of Knowledge Representation and Reasoning", Proceedings of the Seventh International Conference, 2000, San Francisco, CA.

Merali, Y. and Davies, John, "Knowledge capture and utilization in virtual communities", International Conference on Knowledge Capture: Proceedings of the international conference on Knowledge capture, New York, 2001, pp 92–99.

Menzel, C. P., and Mayer, R. J., 1991, IDEF5 Concept Report, Integrated Information Systems Evolution Environment, United States Air Force AL/HRGA, Wright-Patterson Air Force Base, OH.

Miller, Daniel T., "Creating better business outcomes through enterprise integration with advanced building control solutions", IEEE Conference on Control Applications- Proceedings, vol. 2, 1999, pp 1744-1748.

Natalya F. Noy and Deborah L. McGuinness, "Ontology Development 101: A Guide to Creating Your First Ontology", 2002, Stanford University, Stanford, CA.

Neuscheler, F., "The economic View: a concept using benchmark to analyze. In: Evaluate and optimize Business Process", IMSE, 1994, France.

Nonaka, I. and Takeuchi, H., The Knowledge-Creating Company, Oxford University Press, New York, 1995.

Noori, H. and Mavaddat, F., "Enterprise integration: issues and methods", International Journal of Production Research, vol. 36, n. 8, 1998, pp 2083-2097.

Oliveira, E., K. Fischer, *et al.*, "Multi-agent systems: which research for which applications." Robotics and Autonomous Systems, vol. 27, 1999, pp 91-106.

Ong, S. K. and Guo, D. O., "Online design reuse tool for the support of the generation, embodiment and detailed design of products, International journal of production research, 15 august 2004, vol. 42, no. 16, pp 3301–3331.

Opdahl, A. L. and Henderson-Sellers, B., "Evaluating and Improving of modeling languages using the BWW-model," C. N. G. Dampney (Ed.), Sydney, 1999, pp 31-38.

Ortiz, Angel F. Lario and Ros, L., "Enterprise Integration- Business process Integrated Management: a proposal for a methodology to develop Enterprise Integration Programs", Computers in Industry, vol. 40, 1999, pp 155-171.

Palmer, Dean, "Total integration: the next step for manufacturing", Manufacturing Computer Solutions, vol. 7, n. 2, 2001, pp 24-25.

Pena-Mora, F., K. Hussein, *et al.*, "CAIRO: a Concurrent Engineering Meeting Environment for Virtual Design Teams." Artificial Intelligence in Engineering: 2000, pp 202-219.

Perakath C. Benjamin, Christopher P. Menzel, Richard J. Mayer, Florence Fillion, Michael T. Futrell, Paula S. deWitte, and Madhavi Lingineni, "Information Integration for Concurrent Engineering", IDEF5 Method Report, 1994.

Perakath C. Benjamin, Christopher P. Menzel, Richard J. Mayer, Natarajan Padmanaban, "Toward a method for Acquiring CIM Ontologies", International Journal of CIM, vol. 8, n. 3, 1995, pp 225-234.

Pilling, Stephen, "Key characteristics: the key to a robust product design", Engineering Technology, Vol. 7, No. 6, 2004, pp 19- 20.

Priest, John W. and Sanchez, Jose M., *Product Development and Design for Manufacturing*, 2nd edition, Marcel Dekker, Inc., New York, 2001.

Richard J. Mayer, Christopher P. Menzel, and Perakath C. Benjamin, "The role of ontology in enterprise integration", The proceedings of IDEF user group conference, 1993.

Robert Jasper and Mike Uschold, "A Framework for Understanding and Classifying Ontology Applications", The proceedings of the IJCAI ontology workshop, 2000.

Rodgers, Paul A., Nicholas H. M. Caldwell, P. John Clarkson And Avon P. Huxor, "The management of concept design knowledge in modern product development organizations", Int. J. Computer Integrated Manufacturing, Vol. 14, No. 1, 2001, pp 108–115.

Sarder, MD B., "Developing a design ontology for products & processes", PhD dissertation, The University of Texas at Arlington, December, 2006.

314

Sarder, MD B. & Don H. Liles, "Design Ontology Modeling Using IDEF5", Proceedings of the 17th Annual Conference of the Production and Operations Management Society, Boston, Massachusetts, April 28 –May 1, 2006.

Sarder, MD B., Don H. Liles, & Jamie Rogers, "Functional Integration of Manufacturing Enterprises", Proceedings of the Annual Portland International Center for Management of Engineering and Technology Conference (PICMET), Istanbul, Turkey, July 9 - 13, 2006.

Sarder, MD B., John Priest, Don H. Liles, & Young-moon Leem, "Activity Modeling Of Product And Process Design Using IDEF0", Proceedings of the 36th International Conference on Computers & Industrial Engineering, Taipei, Taiwan, June 20 –23, 2006.

Sarder, MD B. & Don H. Liles, "EII – A Quantitative Measure of Functional Integration for Manufacturing Enterprise", Proceedings of the Annual Industrial Engineering Research Conference (IERC), Orlando, Florida, May 20 –24, 2006.

Sarder, MD B., Don H. Liles & John Priest, "Mapping Design Activities Across Product Development Life Cycle", Proceedings of the Annual Industrial Engineering Research Conference (IERC), Orlando, Florida, May 20 –24, 2006.

Sarder, MD B. & Don H. Liles, "The Use of Ontologies for product & process design in Manufacturing Enterprises", Proceedings of the 10th Annual International Conference on Industrial Engineering Theory, Applications & Practice, Clearwater, Florida, December 4 - 7, 2005.

Shen, W. and Norrie, D.H., "An Agent Based Approach for Manufacturing Enterprise Integration and Supply Chain management", Div. of Manufacturing engineering, The University of Calgary, Canada, 2003.

Shanks, G., Tansley, E. and Weber, R., "Using Ontology To Validate Conceptual Models," Communications of ACM, vol. 46, n. 10, 2003, pp 85-89.

Snow, John, "Knowledge captures and re-use", Portals Magazine, June 22, 2004.

Soffer, P., Golany, B., Dori, D. and Wand, Y., "Modeling Off-the-Shelf Information System Requirements. An Ontological Approach", Requirements Engineering, vol. 6, n. 3, 2001, pp 183- 199.

Tham, K.D., "Representation and Reasoning About Costs Using Enterprise Models and ABC", PhD Dissertation, Enterprise Integration Laboratory, Department of Mechanical and Industrial Engineering, University of Toronto, 1999.

True, Mike & Izzi, Carmine, "Collaborative Product Commerce: Creating Value across the Enterprise, Accenture, 2001.

Vernadat, F., "Enterprise Modeling and Integration: Principles and Applications", Chapman & Hall, 1996.

Wand, Y. and Weber, R., "An Ontological Model of an Information System," IEEE Transactions on Software Engineering, vol. 16, n. 11, 1990, pp 1282-1292.

Wand, Y. and Weber, R., "On the Ontological Expressiveness of Information Systems Analysis and Design Grammars," Journal of Information Systems, vol. 3, n. 4, 1993, pp 217-237.

Wand, Y. and Weber, R., "On the Deep Structure of Information Systems," Information Systems Journal, vol. 5, n. 3, 1995, pp.203-223.

Weber, R., Ontological Foundations of Information Systems, Coopers & Lybrand, Melbourne, 1997.

Wei Jerry, and Lee krajewski, "A Model for Comparing Supply Chain Schedule Integration Approaches", International journal of Production, vol. 38, n. 9, 2000, pp 2099-2123.

Yeh I, Karp PD, Noy NF, Altman RB, "Knowledge acquisition, consistency checking and concurrency control for Gene Ontology", Bioinformatics, Vol. 19, n. 2, 2003, pp 241-248.

Yolanda Gil and Jim Blythe "PLANET: A Shareable and Reusable Ontology for Representing Plans", Proceedings of the AAAI 2000 workshop on Representational Issues for Real-world Planning Systems.

Zachman, J., "A Framework for Information Systems Architecture." IBM Systems Journal, vol. 26, n. 3, 1987, pp 276-292.

Zahir Irani, Marinos T., and Peter E.D. Love, "The impact of enterprise application integration on information system lifecycles", Information and Management, vol. 41, 2003, pp 177-187.

INDEX

R
Requirements analysis
Resource management
Robust design
S
Selection design
Site specific ontology
Source material log
Statement pool
Strategic planning
Support service

T
Taxonomy
Technical requirements
Term pool
Terminologies
Test & validation
V
Variant design
W
Work breakdown structure
World-class manufacturers

AUTHOR BIOGRAPHY

MD B. Sarder is an Assistant Professor of Industrial Engineering & Technology at the University of Southern Mississippi. He received his doctorate from The University of Texas at Arlington in 2006. He holds a Master degree in Industrial Engineering from Wichita State University, Kansas and a B. Sc (Eng.) degree in Mechanical/ Industrial & Production Engineering from Bangladesh University of Engineering & Technology (BUET), Dhaka, Bangladesh. Dr. Sarder authored or coauthored more than 32 articles, predominantly on Ontology Modeling/ Knowledge Management. Between 2006 and 2007 he worked as a postdoctoral research fellow at the Systems Engineering Research Center located at the University of Texas at Arlington. He is currently working on developing ontologies in the domain of systems engineering and supply chain management. He has experience working as a supply chain project manager, production and quality control engineer, and technical consultant.

Dr. Sarder has been actively involved with many professional organizations holding various leadership positions and promoted Industrial Engineering involvement. He has developed new curriculum materials for graduate students in the areas of systems engineering and performed various scholarly activities. Due to his academic excellence and outstanding leadership activities, he received many awards, scholarships, and fellowships. This include an NSF scholarship, the Tau Beta Pi Fellowship, the Gilbreth Memorial fellowship, the John Fargher scholarship, the Alpha Pi Mu award of excellence, Volunteer leader scholarship, University Scholar, Who's Who, etc. His research interests include ontology modeling, knowledge management, enterprise integration, supply chain management, medical informatics, continuous process improvement, and systems engineering.